Ace the Technical Pilot Interview

Ace the Technical Pilot Interview

Gary V. Bristow

McGraw-Hill
New York Chicago San Francisco Lisbon London Madrid
Mexico City Milan New Delhi San Juan Seoul
Singapore Sydney Toronto

The McGraw-Hill Companies

Library of Congress Cataloging-in-Publication Data

Bristow, Gary V.
Ace the technical pilot interview / by Gary V. Bristow.
p. cm.
Includes index.
ISBN 0-07-139609-8
1. Airplanes—Piloting—Examinations—Study Guides. I. Title.

TL710.B65 2002
629.132'5216'076—dc21 2002071775

12 13 14 15 16 17 18 19 20 FGR/FGR 0 9 8 7

ISBN 0-07-139609-8

The sponsoring editor for this book was Shelley Carr, the editing supervisor was Daina Penikas, and the production supervisor was Sherri Souffrance. It was set in Century Schoolbook by Wayne Palmer of McGraw-Hill Professional's composition unit, Hightstown, N.J.

Printed and bound by Quebecor World / Fairfield.

Portions of this book were previously published in *Metallurgical Failure Analysis,* © 1993.

This book is printed on recycled, acid-free paper containing a minimum of 50% recycled, de-inked fiber.

Contents

Introduction

In most pilot interviews where a representative of the airline's flight operations department is present, you probably will be asked some technical questions. The scope of the technical questions that may be asked is vast, and obviously, a major part of the decision as to who gets hired and who does not is based on how well you answer these questions. The degree to which you might be examined in this area varies considerably. Some airlines and operators will ask only one or two of the most common questions, whereas others grill the applicant completely with questions of increasing difficulty that cover several areas. Fortunately, though, the interviewer generally is happy if you can answer an adequate percentage of his or her questions.

Research for this book includes feedback from the following airline interviews: United Airlines, Delta, American Airlines, Cathay Pacific, DragonAir, Singapore Airlines, Korean Air, Thai Airways, Air New Zealand, Quantas, Ansett, British Airways, Virgin Atlantic, KLM, Britannia, Air South West, Air Alaska, SAS, Virgin Express, Lufthansa, and British Midland, as well as numerous regional turboprop operators across North America, Europe, the United Kingdom, Southeast Asia, and Australasia. Therefore, this book provides the answers to questions asked by basic light aircraft to heavy jet operators worldwide and is essentially a reference book so that readers can access the answers to specific questions quickly and efficiently.

Readers need to identify the questions that are appropriate for their own interviews. This should be self-evident; i.e., if you are attending an interview about a B737E, you probably will be asked questions about gas-turbine and jet engines and electronic flight instrument systems (EFISs). Likewise, if you are attending an interview concerning light aircraft you probably will be asked questions about piston/propeller engines and mechanical flight instruments. In identifying chapters, subchapters, or individual questions appropriate to your situation, you will drastically reduce your review material. The reference format of this book lends itself well to identifying an individual's expected area of questioning. A word of caution, though, some turboprop operators may ask jet-type questions.

In addition, it is sometimes possible to find out the area of interest or even the exact technical questions asked by an airline before your interview either by questioning previous candidates or by directly or indirectly accessing the pilot grapevine. Pilot associations and unions also can be good distributors of such information, especially from their Web sites.

The questions herein cover main aviation areas, namely, aerodynamics, engines, navigation, and performance, and these areas are further subdivided to help individuals identify their required review. For example, Chapter 2, "Engines," is subdivided into (1) applied laws, (2) piston engines, (3) propellers, (4) jet/gas turbine engines, and (5) engine fire detection and protection. The questions and answers are further arranged in a logical order so that previous and/or following questions and answers provide background to the next question and/or an extension of the preceding question. Thus, when a series of questions are read together, they form an in-depth description of the subject from the basic to the advanced level of the particular subchapter.

Furthermore, to reduce repetition a reference system is used wherever another question and answer are the source of background to or further information for a particular question. In addition, the individual answers are themselves subdivided to denote the short and long answers. Therefore, when answering a question, you can give either the short answer or the in-depth answer. It is left up to you to determine what your interviewer wishes to hear. I suggest that you give the short answer first, and if the interviewer asks you to further explain your answer, then you can give the long answer. If the interviewer wishes you to further explain your answer, then you need to refer to the surrounding answers in that particular subchapter. Fortunately, though, most questions either have only a short answer or the interviewer only wishes to hear the short version. However, some questions only have long, in-depth answers, especially ambiguous questions, which depending on the application of the question could have various applications of the answer. In this case, the interviewer probably is looking to see how well you rationalize your answer as much as the answer itself.

Some answers relate to a generic system, e.g., the autopilot. Therefore, the possibility exists that this generic answer, while being compatible with the design philosophy of most aircraft, might be incompatible with a particular aircraft type. Thus it is left to the good sense of the reader to be aware of this possibility.

Finally, beware of reversal questions. A particular area can have many applications of a particular question, but in this book such variations have not been identified; it is also left to the good sense of the reader to be aware to this possibility.

Abbreviations

AAL	Above aerodrome level
ABC	Auto boost control
AC	Alternating current
ACARS	Aircraft Communications and Reporting System
ACAS	Airborne collision avoidance system
ADC	Air data computer
ADF	Automatic direction finder
ADI	Attitude directional indicator/instrument
ADS-B	Automated Dependent Surveillance–Broadcast
AGC	Automatic gain control
AGL	Above ground level
AH	Artificial horizon
	Alert height
AIC	Aeronautical Information Circular
ALT	Altimeter
A(M)SL	Above (mean) sea level
A of A	Angle of attack
APFDS	Autopilot flight director system
AP	Autopilot
APS	Aircraft prepared for service
APU	Auxiliary power unit
ASDA(R)	Accelerate stop distance available (required)
ASI(R)	Airspeed indicator (reading)
AT	Autotthrust/throttle
ATC	Air traffic control
ATIS	Automatic terminal information service
ATM	Aerodynamic turning moment
	Air traffic management

BDC	Bottom dead center
BEA	British European Airways
BFO	Beat frequency oscillator
BHP	Brake horsepower
C	Celsius
CAA	Civil Aviation Authority
CAS	Calibrated airspeed
CAT	Clear air turbulence
CB	Cumulonimbus clouds
	Circuit breaker
CDI	Course deviation indicator
CDU	Control display unit
CFIT	Controlled flight into terrain
CI	Cost index
CL	Coefficient of lift
CMD	Command
CN	Compass north
C of A	Certificate of airworthiness
C of G, CG	Center of gravity
C of P, CP	Center of pressure
CP	Critical point
CRT	Cathode-ray tube
CSD (U)	Constant speed drive (unit)
CSU	Constant speed unit
CTM	Centrifugal turning moment
CU	Cumulus cloud
CWS	Control wheel steering
DME	Distance-measuring equipment
DALR	Dry adiabatic lapse rate
DI	Directional indicator
DH(A)	Decision height (altitude)
DC	Direct current
EADI	Electronic attitude directional indicator/instrument
EAS	Equivalent airspeed
EAT	Expected approach time

ED(R)	Emergency distance (required)
EFIS	Electronic flight instrument system
EGPWS	Electronic ground proximity warning system
EGT	Exhausted gas temperature
EHSI	Electronic horizontal situation indicator/instrument
ELR	Environmental lapse rate
EMDA(R)	Emergency distance available (required)
EMF	Electromotive force
EPR	Engine pressure ratio
ETOPS	Extended twin operations
FAA	Federal Aviation Administration
FADEC	Full authority digital engine control
FAF	Final approach fix
FANS	Future air navigation systems
FCC	Flight control computer
FCU	Fuel control unit
	Flight control unit
FD(S)	Flight director (system)
FIR	Flight information regions
FL	Flight level
FMC(S)	Flight management computer (system)
FMS	Flight management system
fpm	Feet per minute
GLS	Global landing system
g	Gram
GPS	Global Positioning System
GPWS	Ground proximity warning system
GLONASS	Global Orbiting Navigation Satellite System
GS	Glideslope
HF	High frequency
hPa	Hectopascal
HSI	Horizontal situation indicator/instrument
HUD	Head-up display
HUGS	Head-up guidance system
HWC	Headwind component

IAF	Initial approach fix
IAS	Indicated airspeed
ICAO	International Civil Aviation Organization
IFR	Instrument flight rules
ILS	Instrument landing system
IMC	Instrument meteorologic conditions
INS	Inertia navigation system
IRS	Inertia reference system
ISA	International standard atmosphere
ITCZ	Intertropical convergence zone
IVSI	Inertia/instantaneous vertical speed indicator/instrument
LDA(R)	Landing distance available (required)
LF	Low frequency
LNAV	Lateral navigation
LOC	Localizer
LRC	Long-range cruise
LSS	Local speed of sound
(M)LW	(Maximum) landing weight
(M)ZFW	(Maximum) Zero Fuel Weight
MABH	Minimum approach breakoff height
MAC	Mean aerodynamic chord
MAP	Manifold absolute pressure
	Missed approach point
mbar	Millibar(s)
MCP	Mode control panel
Mcrit	Critical Mach number
MDF	Mach demonstrated maximum flight diving speed
MDH(A)	Minimum decision height (altitude)
MEA	Minimum en route altitude
MEL	Minimum equipment list
MLS	Microwave landing system
MM	Mach meter
MN	Mach number
	Magnetic north
MRC	Maximum range cruise
MSA	Minimum sector altitude

MSL	Mean sea level
MSU	Mode selector unit
NAV	Navigation
NDB	Nondirectional beacon
NOTAMs	Notices to Airmen
NTOFP	Net takeoff flight path
OAT	Outside air temperature
OBI	Omni bearing instrument
OCH(T)	Obstacle clearance height (time)
PAPI	Precision approach path lights
PAR	Precision approach radar
PNR	Point of no return
psi	Pounds per square inch
QDM	Magnetic bearing (radial) to the station
QDR	Magnetic bearing (radial) from the station
QFE	Airfield pressure elevation datum
QNH	Area pressure altitude datum
R	Density
RAS	Rectified airspeed
RAT	Ram air turbine
RBI	Relative bearing indicator/instrument
RMI	Radio magnetic indicator/instrument
R Nav	Area navigation
ROC	Rate of climb
ROD	Rate of descent
RPM	Revolutions per minute
RTO	Rejected takeoff
RVR	Runway visual range
RW	Ramp weight
S	Span
SALR	Saturated adiabatic lapse rate
SAT	Static air temperature

SFC	Specific fuel consumption
SG	Symbol generator
	Specific gravity
SIDs	Standard instrument departures
SL	Sea level
SRA	Secondary radar approach
SSA	Sector safe altitude
SSR	Secondary surveillance radar
STARs	Standard arrivals
SVFR	Special visual flight rules
SWD	Supercooled water droplets
TAF	Terminal aerodrome forecast
TAS	True airspeed
TAT	Total air temperature
TCAS	Traffic collision and avoidance system
TDC	Top dead center
TGT	Total gas temperature
TOD (A/R)	Takeoff distance (available/required)
TOGA	Takeoff go-around
TOR (A/R)	Takeoff run (available/required)
(M)TOW	(Maximum) takeoff weight
TN	True north
TRU	Transformer rectifier unit
TWC	Tailwind component
UHF	Ultrahigh frequency
V	Velocity
*V*1	Decision speed
*V*2	Takeoff safety speed
*V*a	Maneuver speed
*V*app	Approach speed
VASI	Visual approach slope indicator/lights
VAT	Velocity at the threshold
VDF	VHF direction finding
VFR	Visual flight rules
VHF	Very high frequency

VIMD	Minimum drag speed
VIMP	Minimum power speed
VLF	Very low frequency
VMBE	Maximum brake energy speed
VMC	Visual meteorologic conditions
VMC(G/A)	Minimum control speed (on the ground/in the air)
V/MDF	Velocity/Mach maximum demonstrated flight diving speed
V/MMO	Velocity/Mach maximum operating speed
VMU	Minimum demonstrated unstick speed
VNAV	Vertical navigation
VNE	Never exceed speed
VNO	Normal operating speed
V(M)RA	Rough air speed
VOR	VHF omni range
VR	Rotation speed
*V*ref	Reference speed (for approach)
VS	Stall speed
	Vertical speed
*VS*0	Full flap stall speed
*VS*1	Clean stall speed
WAT	Weight-altitude-temperature
WED	Water equivalent depth

Preface

There was one overriding reason for writing this book, namely, to provide pilots with the answers to the technical questions asked at their interview.

When I prepared for a series of airline interviews, I found that many publications would give examples of the technical questions asked, but none gave the all-important answers. Therefore, I found preparing for these interviews difficult and in particular time-consuming because I had to research several textbooks for answers to known questions. Thus I perceived a need for a single book that covered the technical questions that are asked by airlines across North America, Europe, the United Kingdom, Southeast Asia, and Australasia and the all-important answers.

I sincerely hope that this book is helpful in achieving a successful interview result for the reader.

Gary Bristow

Warning: In some answers, suggestion of generic flight techniques and drills are made. Therefore, the possibility exists that this generic suggestion, while being compatible with the design philosophy of most aircraft, may be incompatible with a particular aircraft. Therefore, it is left to the good sense of the reader to be aware of this possibility and to apply the appropriate answer.

Disclaimer: The publisher cannot ensure that the questions asked in your interview are included in this publication or that the interpretations given in the answers are the exact interpretations required by the interviewer.

Chapter 1

Aerodynamics

Forces/Aerofoil

What are the forces acting on an aircraft in flight?

Drag, thrust, lift, and weight.

When thrust and drag are in equilibrium, an aircraft will maintain a steady speed. For an aircraft to accelerate, thrust must exceed the value of drag. When lift and weight are in equilibrium, an aircraft will maintain a steady, level attitude. For an aircraft to climb, lift must exceed the weight of the aircraft.

In a banked turn, weight is a constant, but lift is lost due to the effective reduction in wing span. Therefore, to maintain altitude in a banked turn, the lift value needs to be restored by increasing speed and/or the angle of attack.

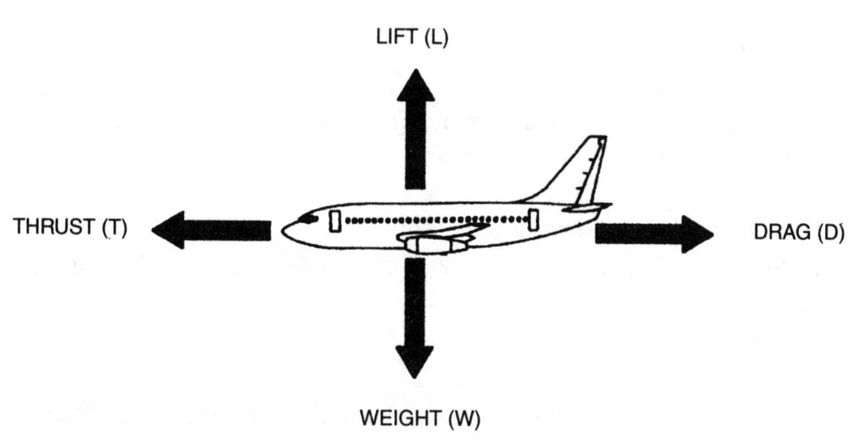

Figure 1.1 Forces acting upon an aircraft in flight.

What produces the maximum glide range?

A maximum lift-drag ratio, obtained by the aircraft being flown at its optimal angle of attack and corresponding minimum drag speed (VIMD), produces an aircraft's maximum glide range.

What is the effect of weight on the glide range?

The glide range does not vary with weight, provided that the aircraft is flown at its optimal angle of attack and speed for that weight, because the glide range is proportional to the lift-drag ratio, which does not vary with weight.

Therefore, if a heavy aircraft were flown at the correct angle of attack and speed, it would glide the same distance as a lighter aircraft. However, the heavier aircraft would have a higher airspeed than the lighter aircraft, and therefore, although it would glide the same distance, it would take less time to do so.

What is rate of climb/descent?

Rate of climb/descent is the vertical component of the velocity of an aircraft and determines the time it will take to either climb or descend from a given height. It is normally expressed in terms of feet per minute.

What is the effect of weight on rate of descent?

The heavier the aircraft, the greater its rate of descent. This is so because a heavy aircraft will fly at a higher airspeed for a given angle of attack, and so its rate of descent will be increased. (*See Qs: How does weight affect an aircraft's flight profile descent point? page 18; Why does an aircraft descend quicker when it is lighter? page 283.*)

What is an aerofoil?

An aerofoil is a body that gives a large lift force compared with its drag when set at a small angle to a moving airstream, e.g., aircraft wings, tailplanes, rudders, and propellers.

What is an aerofoil chord line?

The chord line is a straight line from the leading edge to the trailing edge of an aerofoil.

What is the mean chord line?

The mean chord line is the wing area divided by the wing span (sometimes referred to as the *standard mean chord*).

What is the mean chamberline?

The mean chamberline is a line from the leading edge to the trailing edge of equidistance on the upper and lower surfaces of an aerofoil.

What is the angle of incidence?

The angle of incidence is the angle between the aerofoil's chord line and the aircraft's longitudinal datum. It is a fixed angle for a wing but may be variable for a tailplane. (It is sometimes called *rigging incidence.*)

What is angle of attack?

Angle of attack is the angle between the chord line of an aerofoil and the relative airflow.

What is washout on a wing?

Washout is a decrease in the angle of incidence from the wing root to the tip. This compensates for the early stall due to the higher levels of loading experienced at the wing tips.

What is dihedral?

Dihedral is the upward inclination of a wing from the root to the tip.

What is anhedral?

Anhedral is the downward inclination of a wing from the root to the tip.

Lift

What is lift?

Lift is the phenomenon generated by an aerofoil due to pressure differences above and below the aerofoil.

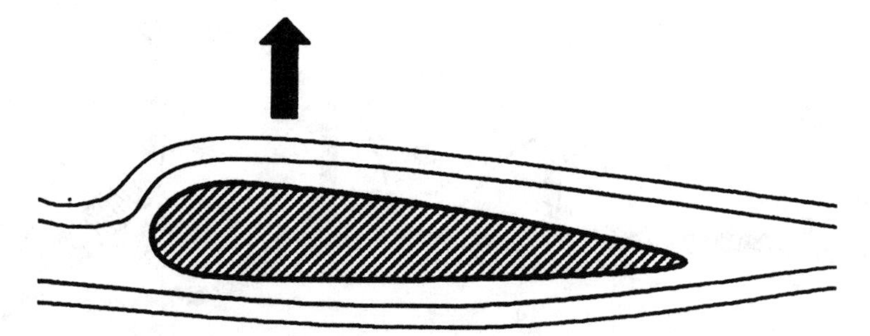

Figure 1.2 Aerofoil-generated lift.

Note: An aerofoil is cambered on its topside and flat on its bottom side. Therefore, the airflow over the top of the aerofoil has to travel farther and thus faster than the airflow below the aerofoil. This causes the pressure below the aerofoil to be greater than above, creating a pressure difference, which results in an upward lift force.

What if the formula for lift?

$$\frac{1}{2}R + V^2 + S + C_L$$

$\frac{1}{2}R$ = half the value of the air density
V^2 = airflow velocity squared
S = wing span area
C_L = coefficient of lift

The combined values of these properties determine the amount of lift produced.

What is coefficient of lift (C_L)?

Coefficient of lift (C_L) is the lifting ability of a particular wing. It depends on both the shape of the wing section (fixed design feature) and the angle of attack.

Describe center of pressure.

The center of pressure is represented as a single point acting on the wing chord line at a right angle to the relative airflow, through which the wing's lifting force is produced.

The position of the center of pressure is not a fixed point but depends on the distribution of pressure along the chord, which itself depends on the angle of attack. Thus, for a greater angle of attack, the point of highest suction (highest air pressure value) moves toward the leading

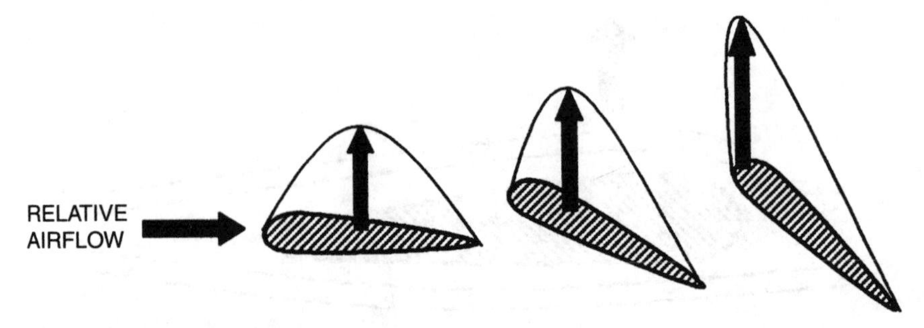

Figure 1.3 Center of pressure position/angle of attack.

edge. The distribution of pressure and center of pressure point thus will be further forward the higher the angle of attack and further aft the lower the angle of attack.

Describe the lift-weight pitching moments.

If the forces of lift and weight are not acting through the same point (line), then they will set up a moment causing either a nose-up or nose-down pitch depending on whether the lift is acting in front of or behind the center of gravity point.

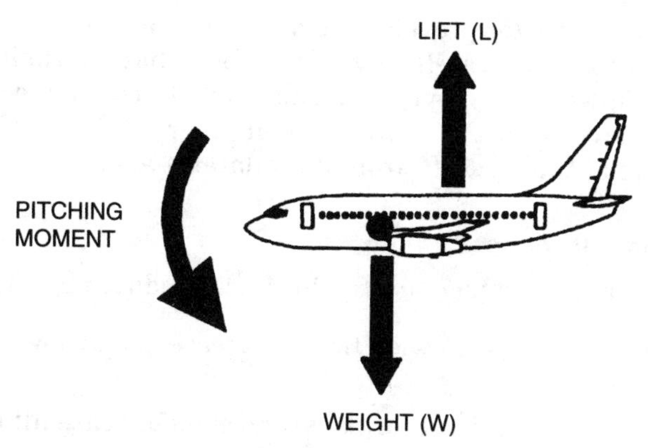

Figure 1.4 Pitching moments.

Note: A center of gravity forward of the center of pitch has a nose-down pitching moment. A center of gravity aft of the center of pitch has a nose-up pitching moment.

The center of pitch moves if the angle of attack changes, and the center of gravity moves as the weight changes (mainly due to fuel being used). Therefore, their positions will vary during a flight.

Describe aspect ratio.

Aspect ratio is the ratio of the wing's span to its geometric chord, e.g., 4:1.
High aspect ratio = high lift (gliders)
Low aspect ratio = lower lift but capable of higher speeds

During what phase of flight is lift the greatest?

In general, the takeoff.

Note: Lift is caused by a pressure difference above and below the wing, and the size of the difference determines the amount of lift produced. (*See Q: What is lift? page 3.*)

The difference in pressure experienced is affected by the functions of lift, which are

1. Configuration (flap setting)
2. Speed of airflow over the wing
3. Angle of attack (which is optimized during the takeoff stage of flight) plus
4. Air density

What is direct lift control?

The elevator/stabilizer provides the direct lift control.

The elevator and stabilizer are aerofoils that by their positions create an upward or downward balancing force that controls the direct lift force from the main aerofoils (wings), thus determining the attitude of the aircraft around the lateral axis.

What are high lift devices?

The following devices increase the lift force produced by the wings:

1. Trailing edge flaps (Fowler flaps) increase lift at lower angles of deflection.
2. Leading edge flaps (Krueger flaps) and slats increase lift by creating a longer wing chord line, chamber, and area.
3. Slots (boundary layer control) prevent/delays the separation of the airflow boundary layer and therefore produce an increase in the coefficient of lift maximum.

Drag

What is drag?

Drag is the resistance to motion of an object (aircraft) through the air.

Define the two major types of drag and their speed relationship.

Profile and induced drag = total drag

Profile drag is also known as *zero-lift drag* and is comprised of

1. Form or pressure drag
2. Skin-friction drag
3. Interference drag

Profile drag increases directly with speed because the faster an aircraft moves through the air, the more air molecules (density) its surfaces encounter, and it is these molecules that resist the motion of the aircraft through the air. This is known as *profile drag* and is greatest at high speeds.

Induced drag is caused by creating lift with a high angle of attack that exposes more of the aircraft's surface to the relative airflow and is associated with wing-tip vortices. A function of lift is speed, and therefore, induced drag is indirectly related to speed, or rather the lack of speed. Thus induced drag is greatest at lower speeds due to the high angles of attack required to maintain the necessary lift. Induced drag reduces as speed increases because the lower angles of incidence associated with higher speeds create smaller wing-tip trailing vortices that have a lower value of energy loss.

Minimum drag speed (VIMD) is the speed at which induced and profile drag values are equal. It is also the speed that has the lowest total drag penalty, i.e.,

$$\text{VIMD} = \text{minimum drag speed}$$

Therefore, this speed also represents the best lift-drag ratio (best aerodynamic efficiency) that will provide the maximum endurance of the aircraft.

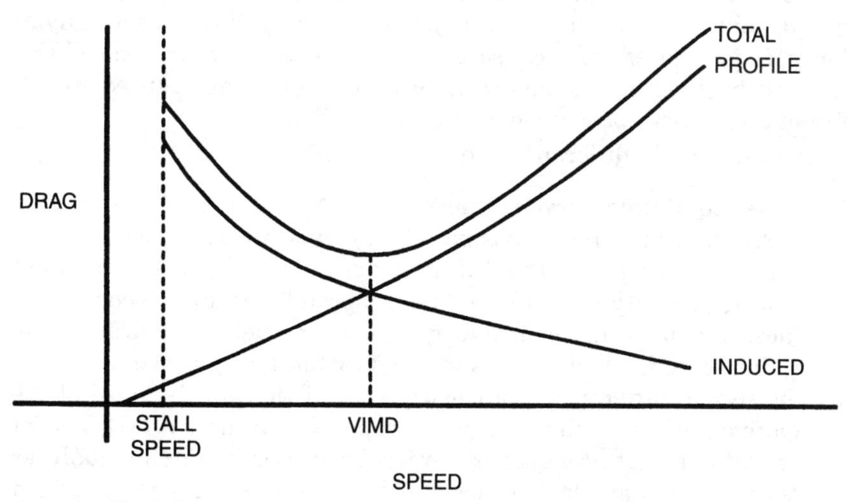

Figure 1.5 Total drag curve.

Describe the drag curve for a piston/propeller aircraft.

For a piston-engined propeller aircraft read straight-winged. It has a typical total drag curve comprised of a well-defined steep profile drag curve at high speeds. This is so because the wing is not designed for high speeds, and therefore, as speed increases, profile drag increases as a direct result. It also has a well-defined induced drag curve at low speeds. This is so because the straight-winged aircraft has a higher C_L value, and with induced drag being proportional to lift, the lower the speed, the greater is the angle of attack required to achieve the necessary lift, and therefore, the greater is the associated induced drag component. It also has a well-defined bottom VIMD (minimum drag speed) point and is capable of a lower stall speed than a jet.

Flight below VIMD in a piston-engined aircraft is very well defined by the steep increase in the drag curve in flight as well as on paper. Speed is not stable below VIMD, and because of the steep increase of the curve below VIMD, it is very noticeable when you are below VIMD. That is, below VIMD, a decrease in speed leads to an increase in drag that causes a further decrease in speed. (*See Q: Explain speed stability. page 22; see also Fig. 1.5, Total Drag Curve, page 7.*)

Describe the drag curve on a jet aircraft.

The drag curve on a jet aircraft is the same as for a a piston aircraft in that it is comprised of induced drag, profile drag, and a VIMD speed, but its speed-to-drag relationship is different. This is so because the jet aircraft has swept wings, which are designed to achieve high cruise speeds, but as a consequence has poorer lift capabilities, especially at low speeds. Therefore, because profile drag is a function of speed and induced drag is proportional to lift, the drag values against speed are different on a jet/swept-winged aircraft.

The three main differences are

1. Flatter total drag curve because
   a. Profile drag is reduced especially against higher speeds.
   b. Induced drag is reduced (flatter drag curve) because the swept wing has very poor lift qualities, especially at low speeds.
   These factors combined give rise to a smaller total drag range against speed, which results in a flatter total drag curve.
2. The second difference is a consequence of the first because of the relative flatness of the drag curve, especially around VIMD. The jet aircraft does not produce any noticeable changes in flying qualities other than a vague lack of speed stability, unlike the piston-engined aircraft, in which there is a marked speed-drag difference. (Speed is unstable below VIMD, where an increase in thrust has a greater drag penalty for speed gained, thus with a net result of losing speed for a given increase in thrust.)
3. VIMD is a higher speed on a jet aircraft because the swept wing is more efficient against profile drag, and therefore, the minimum drag speed is typically a higher value.

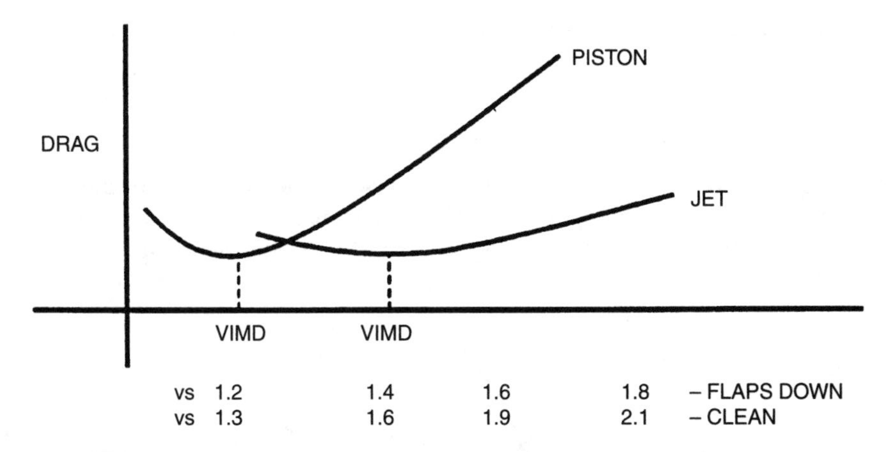

Figure 1.6 Total jet drag curve against the piston engine drag curve.

Describe the pitching moment associated with the thrust-drag couple.

If the forces of thrust and drag are not acting through the same point (line), then they will set up a moment causing either a nose-up or nose-down pitch depending on whether the thrust is acting above or below the dragline.

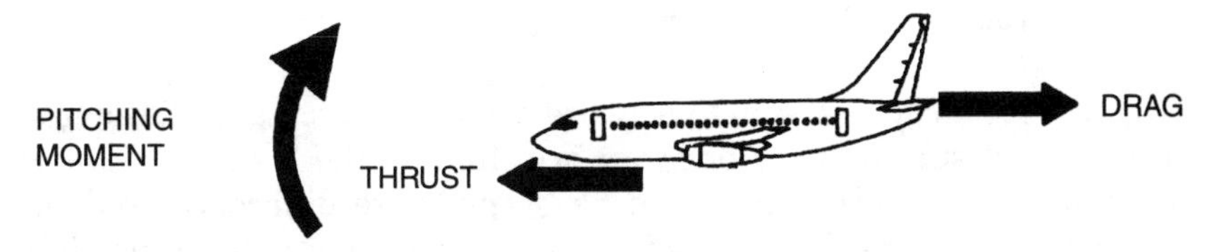

Figure 1.7 Thrust-drag couple pitching moments.

Therefore, a change in thrust (increase or decrease) in straight and level flight can lead to a pitching tendency of the aircraft. Likewise, an increase or decrease in drag also can lead to a pitching tendency of the aircraft. For example, an increase in thrust on an aircraft with engines mounted under the wing, with a higher dragline, will cause a nose-up pitch as thrust is increased.

What are high-drag devices?

The following devices increase the drag penalty on an aircraft:

1. Trailing edge flaps (in high-drag/low-lift position)
2. Spoilers
    a. In fight detent, used as a speed brake
    b. On the ground, used as lift dumpers
3. Landing gear

4. Reverse thrust (ground use only)
5. Braking parachute

What causes/are wing-tip vortices?

Wing-tip vortices are created by span-wise airflow over the upper and lower surfaces of a wing/aerofoil that meet at the wing tips as turbulence and therefore induces drag, especially on a swept wing.

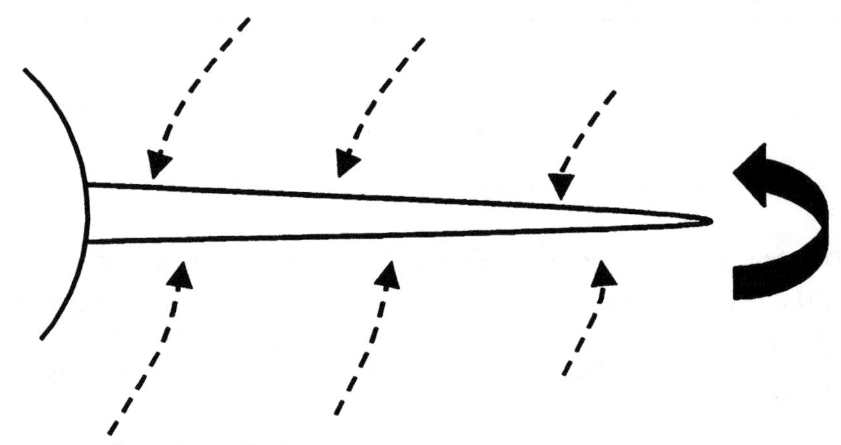

Figure 1.8 Span-wise airflow/vortices on the wing.

Span-wise airflow is created because a wing producing lift has a lower static pressure on the upper surface than on the lower surface. At the wing tip, however, there can be no pressure difference, and the pressure is equalized by air flowing around the wing tip from the higher pressure on the lower surface to the lower pressure on the upper surface. There is therefore a span-wise pressure gradient, i.e., pressure changing along the wing span.

What are the effects of span-wise airflow over a wing?

1. Creates wing-tip vortices.
2. Reduced aileron (wing control surface) efficiency.
3. Reversed span-wise airflow increases disturbed airflow on the wing's upper surface at the tip, contributing to a wing-tip stall.

What are the effects of wing-tip vortices?

1. Create aircraft drag (induced drag because the vortices induce a downward velocity in the airflow over the wing, causing a change

in the direction of the lift force so that it has an induced-drag component; therefore, it creates a loss of energy).

2. Vortices create turbulence, which may affect the safety of other aircraft within approximately 1000 ft below or behind the aircraft.
3. Downwash affects the direction of the relative airflow over the tailplane, which affects the longitudinal stability of the aircraft.

How do you prevent span-wise airflow on a wing, especially a swept wing?

Fences and vortex generators. These items direct the airflow over the wing's upper surface perpendicular to the leading edge.

What is the purpose of vortex generators/fences?

To reduce span-wise airflow and thereby reduce its effects.

One of the effects of span-wise airflow over the wing is reduced effectiveness of the ailerons due to the diagonal airflow over the control surfaces. Vortex generators are located on the upper surface of a wing to create a slightly disturbed airflow perpendicular to the leading edge of the wing, which helps to maximize the effectiveness of the control surfaces, especially the ailerons.

Fences also help to maximize the effectiveness of the control surfaces in a similar yet cruder manner. However, they are used normally to reduce the reverse span-wise airflow on the upper wing surface from reaching the wing tips, thus reducing the airflow, which contributes to a wing-tip stall.

Vortex generators are used in other areas of the aircraft where a disturbed airflow is required (disturbed air tends to be denser and of a slower velocity), such as inlets to some types of auxiliary power units (APUs).

What are winglets, and how do they work?

Winglets are aerodynamically efficient surfaces located at the wing tips. They are designed to reduce induced drag. They dispense the span-wise airflow from the upper and lower surfaces at different points, thus preventing the intermixing of these airflows that otherwise would create induced-drag vortices.

Weight and Aircraft Momentum

For aircraft weight definitions, see Chapter 7, "Performance," subchapter, "Loading."

What limits an aircraft's structural weight?

The main force generated to balance the aircraft's gross weight is the lift force, and if the lift cannot equal the aircraft's weight, then the aircraft cannot maintain level flight. Therefore, the aircraft weight is directly restricted by the lift capabilities of the aircraft.

Note: The lift force generated is limited by the size (design) of the wing, the attainable airspeed (airspeed is limited by the power available from the engine/propeller), and the air density.

What are the effects of excessive aircraft weight?

If the limiting weight of an aircraft is exceeded, the following effects are experienced:

1. Performance is reduced.
   a. Takeoff and landing distance is increased.
   b. Rate of climb and ceiling height are reduced.
   c. Range and endurance will be reduced.
   d. Maximum speed is reduced.
2. Stalling speed is increased.
3. Maneuverability is reduced.
4. Wear on tires and brakes is increased.
5. Structural safety margins are reduced.

Describe center of gravity.

The center of gravity (C of G, CG) is the point through which the total weight of a body will act.

Describe a component arm.

The definition of a component arm is the distance from the datum to the point at which the weight of a component acts (center of gravity point). By convention, an arm aft of the datum, which gives a nose-up moment, is positive, and an arm forward of the datum, which gives a nose-down moment, is negative. Therefore, for a constant weight, the longer the arm, the greater is the moment.

Describe center of gravity moment.

The moment is the turning effect/force of a weight around the datum. It is the product of the weight multiplied by the arm:

$$\text{Moment} = \text{weight} \times \text{arm}$$

How is the pitching moment of the lift-weight couple balanced?

When the pitching moment of the lift-weight couple is not balanced perfectly (*see Q: Describe the lift-weight pitching moments? page 3*), extra forces are provided by the horizontal tailplane to center the aircraft's pitching moment.

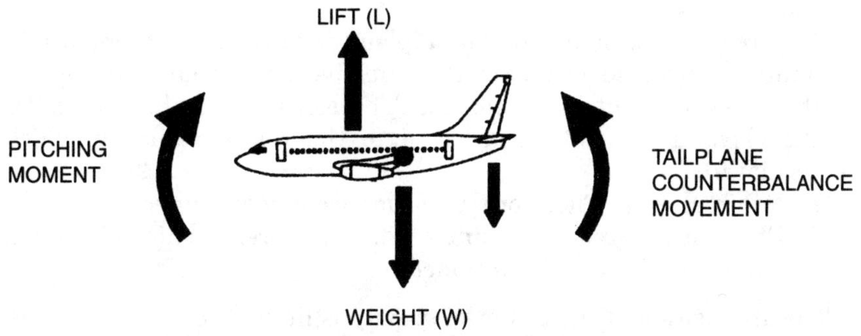

Figure 1.9 Pitching moment balanced by the horizontal tailplane.

Note: Lift forward of weight has a nose-up pitching moment, which is counterbalanced by the downward deflection of the horizontal tailplane, which creates a nose-down counterpitch. Therefore, lift aft of weight requires the opposite balance.

The tailplane force has a turning moment in the pitching plane (nose up or nose down) about the lateral axis at the center of gravity point. Its effectiveness depends on its size and the length of its moment arm from the center of gravity.

Describe the center of gravity range.

The center of gravity range relates to the furthest forward and aft center of gravity positions along the aircraft's longitudinal axis, inside which the aircraft is permitted to fly. This is so because the horizontal tailplane can generate a sufficient lift force to balance the aircraft's lift-weight moment couple so that it remains longitudinally stable and retains a manageable pitch control. (Center of gravity range or envelope is listed in the aircraft's flight manual, and accordance is mandatory.) (*See Q: What is longitudinal stability? page 33.*)

What are the reasons/effects of keeping a center of gravity inside its limits?

The forward position of the center of gravity is limited to

1. Ensure that the aircraft is not too nose heavy so that the horizontal tailplane has a sufficient turning moment available to overcome its natural longitudinal stability.

2. Ensure that the aircraft's pitch control (rotation and flare) is not compromised, with high stick forces (tailplane turning moment), by restricting the aircraft's tailplane arm forward center of gravity limit. [Remember, tailplane moment (stick force) = arm × weight.] Note that this is particularly important at low speeds (i.e., takeoff and landing), when the elevator control surface is less effective.
3. Ensure a minimum horizontal tailplane deflection, which produces a minimal download airforce on the tailplane and is required to balance the lift-weight pitching moment. Therefore, the stabilizer and/or the elevator is kept streamlined to the relative airflow, which results in
   a. Minimal drag. Therefore, performance is maintained.
   b. Elevator range being maintained. Therefore, the aircraft's pitch maneuverability is maintained.

The aft position of the center of gravity is limited to

1. Ensure that the aircraft is not too tail heavy so that the horizontal tailplane has a sufficient turning moment available to make the aircraft longitudinally stable.
2. Ensure that enough pitch control stick forces (tailplane turning moment) are adequately felt through the control column by guaranteeing the aircraft's tailplane arm to an aft center of gravity limit. [Remember, moment (stick force) = arm × weight.]
3. Ensure a minimum horizontal tailplane deflection, which produces a minimal upload airforce on the tailplane and is required to balance the lift-weight pitching moment. Therefore, the stabilizer and/or the elevator is kept streamlined to the relative airflow, which results in
   a. Minimal drag. Therefore, performance is maintained.
   b. Elevator range being maintained. Therefore, the aircraft's pitch maneuverability is maintained.

What are the effects of a center of gravity outside its limits (range)?

If the center of gravity is outside its forward limit, the aircraft will be nose heavy, and the horizontal tailplane will have a long moment arm (tailpipe to center of gravity point) that results in the following:

1. Longitudinal stability is increased because the aircraft is nose heavy. (*See Q: What is longitudinal stability? page 33.*)
2. The aircraft's pitch control (rotation and flare) is reduced or compromised because it experiences high stick forces due to the aircraft's long tailplane moment arm. [Remember, tailplane moment (stick force) = arm × weight.]

3. A large balancing download is necessary from the horizontal tailplane by deflecting the elevator or stabilizer. This results in

   a. An increased wing angle of attack resulting in higher induced drag, which reduces the aircraft's overall performance and range.
   b. Increased stalling speed due to the balancing download on the horizontal tailplane, which increases the aircraft's effective weight. (*See Q: How does a forward center of gravity affect the stall speed, and why? page 16.*)
   c. Also, if the elevator is required for balance trim, less elevator is available for pitch control, and therefore, the maneuverability of the aircraft to rotate at takeoff or to flare on landing is reduced.
   d. In-flight minimum speeds are also restricted due to the lack of elevator available to obtain the necessary high angles of attack required at low speeds.

Generally, the aircraft is heavy and less responsive to handle in flight and requires larger and heavier control forces for takeoff and landing.

If the center of gravity is outside its aft limit, the aircraft will be tail heavy, and the horizontal tailplane will have a short moment arm (tailplane to center of gravity point) that results in the following:

1. The aircraft is longitudinally unstable because it is too tail heavy for the horizontal tailplane turning moment to balance. (*See Q: What is longitudinal stability? page 33.*)
2. The aircraft's pitch control (rotation and flare) is increased (more responsive) because it experiences light stick forces due to the aircraft's short tailplane arm. [Remember, tailplane moment (stick force) = arm × weight.] This lends itself to the possibility of overstressing the aircraft by applying excessive *g* forces.
3. A large balancing upload is necessary from the horizontal tailplane by deflecting the elevator or stabilizer. This results in

   a. A decreased wing angle of attack, resulting in lower induced drag, which increases the aircraft's overall performance and range.
   b. A lower stalling speed due to the balancing upload on the horizontal tailplane, which decreases the aircraft's effective weight.
   c. Also, if the elevator is required for balance trim, less elevator is available for pitch control, and therefore, the maneuverability of the aircraft to recover from a pitch-up stall attitude is reduced.
   d. In-flight maximum speeds are also restricted due to the lack of elevator available to obtain the necessary low angles of attack required at high speeds.

Generally, the aircraft is effectively lighter and more responsive to handle in flight and requires smaller and lighter control forces for takeoff and landing.

Summary

Center of Gravity Position	Forward	Aft
Stability	More	Less
Stick force	More	Less
Drag	More	Less

What factors determine the loading (weight and balance) of an aircraft?

(*See Chapter 7, "Performance," subchapter, "Loading," page 184.*)

If you were loading an aircraft to obtain maximum range, would you load it with a forward or aft center of gravity (forward or aft cargo hold)?

An aft center of gravity position/hold loading, for aircraft (especially jet/swept wing) with a nose-up en route attitude will allow it to achieve its maximum possible range.

An aft center of gravity position, normally is accomplished by using the aft cargo hold, which gives the aircraft its nose-up en route attitude naturally; therefore, the stabilizer can remain streamlined to the airflow and produce no relevant drag (e.g., aft center of gravity = 6° nose-up attitude = 0° elevator/stabilizer deflection). Thus the aircraft can be operated at its optimal thrust setting to obtain its maximum range without having to use excessive engine thrust to compensate for drag.

Note: It is beneficial even when the center of gravity is aft of its optimal position because the stabilizer would produce a greater lift force (to produce a downward pitching moment of the nose to gain its en route attitude). This is beneficial to the aircraft's overall performance because it increases the aircraft's overall lift capabilities, whereas a forward center of gravity has a detrimental effect on the aircraft's performance. However, a few aircraft with a nose-down en route attitude would require a forward center of gravity position.

How does a forward center of gravity affect the stall speed, and why?

A center of gravity forward of the center of pressure will cause a higher stall speed.

This is so because a forward center of gravity would cause a natural nose-down attitude below the required en route cruise attitude for best performance. Therefore, a downward force is induced by the stabilizer to obtain the aircraft's required attitude. However, this downward force is in effect a weight and thereby increases the aircraft's overall effective weight. Weight is a factor of the stall speed, and the heavier the aircraft, the higher is the aircraft's stall speed.

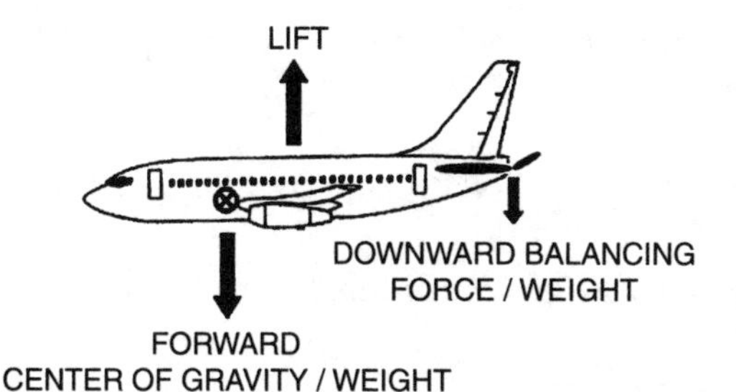

Figure 1.10 Forward center of gravity position effect on stall speed.

Conversely, the opposite is true: A center of gravity aft of the center of pressure will cause a lower stall speed.

Why does a jet aircraft have a large center of gravity range?

A jet aircraft needs a large center of gravity range because its center of gravity position can change dramatically with a large change in its weight during a flight. (*See Q: What causes center of gravity movement? page 17.*) Therefore, to accommodate a large center of gravity movement, the aircraft has to have a powerful horizontal tailplane to balance the large lift-weight pitching moments so that the aircraft remains longitudinally stable and retains its pitch controllability. (*See Q: What are the four reasons for a variable incidence tailplane? page 43.*)

What causes center of gravity movement?

The center of gravity is the point through which weight acts. Therefore, movement of the center of gravity is due to a change in weight.

The distribution of the aircraft's weight can change for three reasons and thus cause the center of gravity position to move. The three reasons for a change in the aircraft's weight are

1. *Fuel burn.* The most common reason for center of gravity movement on a swept-wing aircraft is its decrease in weight as fuel is used in flight. It should be remembered that because of the sweep, the wing and the fuel tanks housed inside cover a distance along the aircraft's longitudinal axis. Therefore, as fuel/weight is reduced progressively along this axis, the weight-distribution pattern changes across the aircraft's length.
2. *Passenger movement.*
3. *High speeds.*

Note: This is so because the greater the speed, the greater is the lift created. To maintain a straight and level attitude, the aircraft adopts a more nose-down profile, which is accomplished by creating lift at the tailplane. This lift on the tailplane effectively reduces the weight of the tailplane section of the aircraft.

How does minimum control speed in the air (VMCG/A) vary with center of gravity position?

(*See Chapter 7, "Performance and Flight Planning," page 195.*)

Describe the effects of an aircraft's momentum.

Momentum of a body is the product of the mass of the body and its velocity, which enables the body (aircraft) to remain on its previous direction and magnitude for a quantity of time after an opposing force has been applied.

The momentum of a jet aircraft is significantly greater than that of a piston-engined aircraft in all its handling maneuvers, climbs, descents, and turns because of its greater weight and velocity.

How does weight affect an aircraft's flight profile descent point?

The heavier the aircraft, the earlier is its required descent point.

The heavier the aircraft, the greater is its momentum, remember, momentum = mass × velocity. Therefore, for a constant indicated air speed (IAS) or Mach number (i.e., its V/MMO), the heavier aircraft will have to maintain a shallower rate of descent to check its momentum. (*See Qs: What is the effect of weight on rate of descent? page 2; Why does an aircraft descend quicker when it's lighter? page 283.*)

The shallower the rate of descent, the greater is the ground speed, and because an aircraft's descent is a function of rate of descent (ROD), the aircraft will cover a greater distance over the ground per 100-ft descent or per minute.

Therefore, total descent is measured against distance over the ground, which is a function of ground speed, which depends on momentum, which depends on weight. Thus the greater the aircraft's weight, the earlier is its required descent point.

What is positive *g* force?

Positive *g* force is the influence of the force of gravity on the normal terrestrial environment beneath it. This is perceived as the normal weight of any body, including ourselves, in the terrestrial environment, i.e., 1*g*.

What is negative *g* force?

Negative *g* force is the opposite to positive *g* force (i.e., the influence of the normal terrestrial environment above the force of gravity, 1*g*).

Swept Wing

Describe how you would design a high-speed aircraft wing.

Thin, minimal-chamber, swept wings.

In designing a high-speed wing, you need to consider first the requirement for economical high-speed performance in the cruise configuration. However, you also have to consider the restraints on the design of the need to keep airfield performance within acceptable limits and the need to give the structural people a reasonable task.

There are several interactive design areas of a wing. Some are purely for lift, some are a compromise between lift and speed, and some are purely for speed. For the high-speed requirements of a wing, the design would focus on *sweep, thickness,* and *chamber.*

The degree of sweep, thickness, and chamber used for the final high-speed wing design depends on their many interactive compromises that culminate in directly fulfilling the wing's high-speed requirement and inversely the wing's lift and structural requirements.

How does a swept wing aid the increase in its critical Mach number (M_{crit}) speed?

The swept-wing design increases its M_{crit} speed because it is sensitive to the (airflow) airspeed vector normal to the leading edge for a given aircraft Mach number. A swept wing makes the velocity vector normal (perpendicular) (*AC*) to the leading edge a shorter distance than the chordwise resultant (*AB*). Since the wing is responsive only to

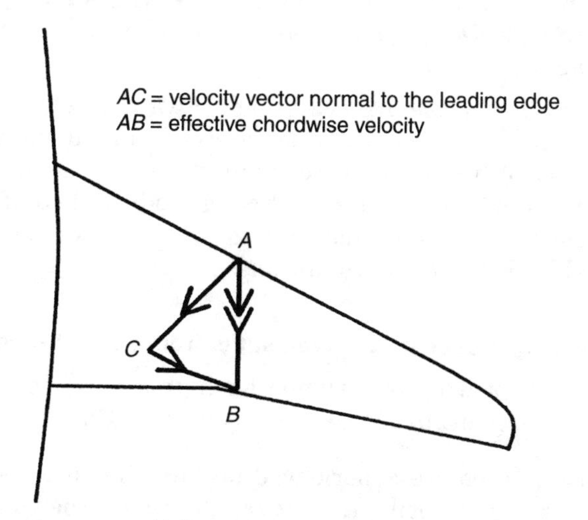

Figure 1.11 Airflow velocity vectors over a swept wing.

the velocity vector normal to the leading edge, the effective chordwise velocity is reduced (in effect, the wing is persuaded to believe that it is flying slower than it actually is). This means that the airspeed can be increased before the effective chordwise component becomes sonic, and thus the critical number is raised.

Describe how you would optimize the lift design on a swept wing.

To optimize the lift design on a swept wing, you would need to (1) examine and develop the lift design areas of the clean wing and (2) add high lift devices to the clean wing to a degree that satisfies our main lift concern, that of adequate airfield performance.

What advantages does a jet aircraft gain from a swept wing?

The advantages a jet aircraft gains from the swept wing are (1) high Mach cruise speeds and (2) stability in turbulence.

1. *High Mach cruise speeds.* The swept wing is designed to enable the aircraft to maximize the high Mach speeds its jet engines can produce. The straight-winged aircraft experiences sonic disturbed airflow, resulting in a loss of lift at relatively low speeds. Therefore, a different wing design was required for the aircraft to be able achieve higher cruise speeds. The swept-wing design delays the airflow over the wing from going supersonic and, as such, allows the aircraft to maximize the jet engine's potential for higher Mach cruise speeds. Additionally, the swept wing is also designed with a minimal chamber and thickness, thereby reducing profile drag, which further increases the wing's ability for higher speeds. (*See Q: Describe how you would design a high-speed aircraft wing, page 19.*)
2. *Stability in turbulence.* Ironically, a disadvantage of the swept wing is its poor lift qualities, which lends itself to an advantage in that it is more stable in turbulence compared with a straight-winged aircraft. This is so because the swept wing produces less lift and therefore is less responsive to updraughts, which allows for a smoother, more stable ride in gusty conditions.

What disadvantages does a jet aircraft suffer from a swept wing?

Because the swept wing is designed for high cruise speeds, it suffers from the following disadvantages as a consequence:

1. Poor lift qualities are experienced because the sweep-back design has the effect of reducing the lift capabilities of the wing.

2. Higher stall speeds are a consequence of the poor lift qualities of a swept wing.

3. Speed instability is the second consequence of poor lift at lower speeds for the swept-wing aircraft. Speed is unstable below minimum drag speed (VIMD) because the aircraft is now sliding up the back end of the jet drag curve, where power required increases with reducing speed. This means that despite the higher coefficient of lift (C_L) associated with lower speeds, the drag penalty increases faster than the lift; therefore, the lift-drag ratio degrades, and the net result is a tendency to progressively lose speed. Thus speed is unstable because of the drag penalties particular to the swept wing. (*See Q: Explain speed stability, page 22.*)

4. A wing-tip stalling tendency is particular to a swept-wing aircraft mainly because of the high local C_L loading it experiences. Uncorrected (in the design), this effect would make the aircraft longitudinally unstable, which is a major disadvantage. (*See Q: Where does a swept wing stall first, and what effect does this have on the aircraft's attitude? page 21.*)

Where does a swept wing stall first, and what effect does this have on the aircraft's attitude?

A simple swept and/or tapered wing will stall at the wing tip first if not induced/controlled to stall at another wing section first by the designer.

This is so because the outer wing section experiences a higher aerodynamic loading due to the wing taper, which causes a greater angle of incidence to be experienced to a degree where the airflow stalls at the wing tips. The boundary layer span-wise airflow, also a result of sweep, further contributes to the airflow stalling at the wing tips.

A stall at the wing tip causes a loss of lift outboard and therefore aft (due to the wing sweep), which moves the center of pressure inboard and therefore forward; this produces a pitch-up tendency that continues as the wing stalls progressively further inboard.

A wing-tip stall is resolved in the wing design with the following better aerodynamic stalling characteristics:

1. Greater chamber at the tip; this increases airflow speed over the surface, which delays the stall.

2. Washout or twist, which creates a lower angle of incidence at the wing tips and delays the effect of the outboard wing loading that causes the stall.

Speed

For performance speed definitions, see Chapter 7, "Performance and Flight Planning." For airspeed definitions, see Chapter 5, "Atmosphere and Speed."

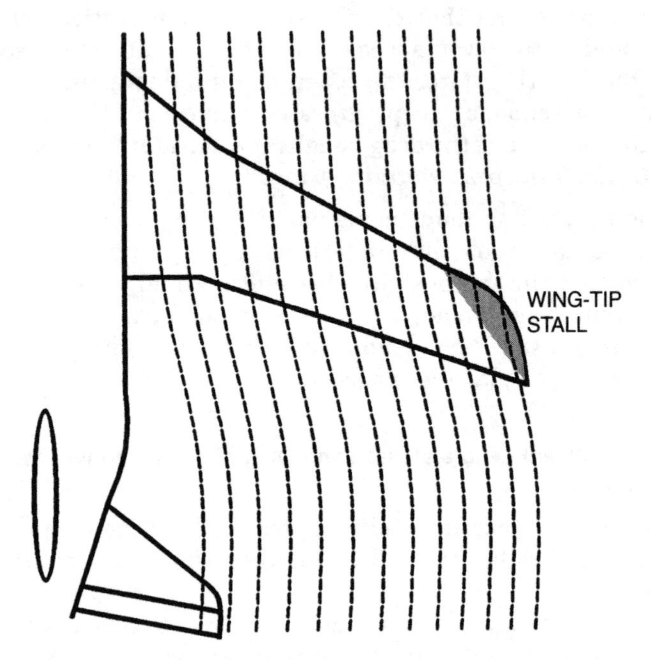

Figure 1.12 Wing-tip stall.

Explain speed stability.

Speed stability is the behavior of the speed after a disturbance at a fixed power setting.

The behavior of an aircraft's speed after it has been disturbed is a consequence of the drag values experienced by the aircraft frame.

Speed is said to be stable if after it has been disturbed from its trimmed state it returns naturally to its original speed. For example:

1. An increase in speeds leads to an increase in drag, thus causing a return to the original speed.
2. A decrease in speed leads to a decrease in drag, thus causing a return to the original speed.

Speed is said to be unstable if after it has been disturbed from its trimmed state the speed divergence continues, resulting in a negative speed stability. For example:

1. A decrease in speed leads to an increase in drag, which causes a further decrease in speed, thus causing a negative speed divergence.
2. An increase in speed leads to a decrease in drag, which causes a further increase in speed, thus causing a positive speed divergence.

What is Mach number?

Mach number (MN) is a true airspeed indication, given as a percentage relative to the local speed of sound; e.g., half the speed of sound = 0.5 Mach. (*See Q: Describe Mach number, page 122.*)

What is the critical Mach number (M_{crit})?

M_{crit} is the aircraft's Mach speed at which the airflow over a wing becomes sonic—critical Mach number.

The aircraft's Mach speed is lower than the airflow speed over a wing. A typical M_{crit} speed of 0.72 M experiences sonic Mach 1 airflow speed over the upper surface of the wing. Subsonic aircraft experience a rapid rise in drag above the critical Mach number, and because the aircraft's engines do not have the available power to maintain its speed and lift values under these conditions, the aircraft suffers a loss of lift.

Describe the characteristics of critical Mach number (M_{crit})?

1. Initial Mach buffet, caused by the shock waves on the upper surface of the wing as the aircraft approaches M_{crit}, is usually experienced.
2. An increase in drag because of the breakdown of airflow causes the stick force to change from a required forward push to a neutral force and then a required pull force as the aircraft approaches and passes M_{crit}.
3. A nose-down change in attitude (Mach tuck) occurs at or after M_{crit}.
4. A possible loss of control.

Describe the changes in the center of pressure as an aircraft speed increases past the critical Mach number (M_{crit}).

The center of pressure moves rearward on a swept wing as the aircraft passes its M_{crit} for two reasons:

1. The shock waves on the wing's upper surface occur toward the leading edge because of the greater chamber, which creates the greatest airflow velocity to be experienced at this point. This upsets the lift distribution chordwise and causes a rearward shift in the center of the lift (center of pressure).
2. The swept wing tends to experience the shock-wave effect at the thick root part of the wing first, causing a loss of lift inboard, and therefore, the lift force now predominantly comes from the outboard part of the swept wing, which is further aft because of the wing sweep. (*See Q: Describe center of pressure, page 4.*)

What is Mach tuck?

Mach tuck is the nose-down pitching moment an aircraft experiences as it passes its critical Mach number (M_{crit}).

Mach tuck is a form of longitudinal instability that occurs because of the center of pressure's rearward movement behind the center of gravity (see preceding question), which induces the aircraft to pitch down (or the aircraft's nose to tuck).

What is the purpose of a Mach trimmer?

The purpose of a Mach trimmer is to automatically compensate for Mach tuck (longitudinal instability) above M_{crit}. (*See Q: What is Mach tuck? page 24.*)

What is a Mach trimmer, and what is it used for?

A Mach trimmer is a system that artificially corrects for Mach tuck above the aircraft's M_{crit} by sensing the aircraft's speed and signaling a proportional upward movement of the elevator or variable-incidence stabilizer to maintain the aircraft's pitch attitude throughout its speed range up to its maximum Mach demonstrated flight diving speed (MDF).

Note: Mach trimmers allow for an aircraft's normal operating speed range to be above its M_{crit}.

In the event of a Mach trimmer failure, there is usually an imposed reduced Mach maximum operating speed (MMO) value so that a margin is retained below the Mach speed at which the onset of instability occurs.

What are the effects of compressibility?

Compressibility is the effect of air being compressed onto a surface (at a right angle to the relative airflow), resulting in an increase in density, and thus dynamic pressure rises above its expected value. It is directly associated with high speeds. (*See Chapter 5, "Atmosphere and Speed," page 115.*)

There are two main effects of compressibility:

1. Compressibility error on dynamic pressure reading flight instruments; e.g., air speed indicator shows an overread error that is greater the higher the aircraft's speed. (*See Q: Describe equivalent air speed [EAS], page 121.*)
2. Compressed air is experienced on the leading edge of the wing, which disturbs the pressure pattern on the wing and causes the disturbed air shock-wave/drag effect at the critical Mach number. (*See Q: What is the critical Mach number [M_{crit}]? page 23.*)

What is the main influence on Mach number?

(*See Chapter 5, "Atmosphere and Speed," page 123.*)

Explain speed margins.

A speed margin is the difference between the aircraft's normal maximum permitted operating speed and its higher certified testing speed.

For a piston-engined propeller aircraft:

VNO is the normal operating maximum permitted speed.

VNE is the higher, never exceeded operating speed.

VDF is the maximum demonstrated flight diving speed, established during design certification flight trials.

The piston-engined propeller aircraft enjoys a relatively large margin between VNO and VDF and has very little overspeed tendencies. Therefore, the speed margin for a piston-engined propeller aircraft is not very significant.

For the jet aircraft:

VMO/MMO is the maximum indicated operating speed in knots or Mach number. This is the normal maximum operating speed, which ensures an aircraft's structural integrity and adequate handling qualities.

VDF/MDF is the maximum demonstrated flight diving speed in knots or Mach number established during the design certification flight trials. This flight diving speed incurs reduced aircraft structural integrity and often a lower level of handling qualities.

The jet aircraft's margin between VMO/MMO and VDF/MDF is relatively small, and because of its low cruise drag and the enormous power available from its jet engines, especially at low altitudes, the jet aircraft has a distinct overspeed tendency. Therefore, the speed margin on a jet aircraft is very significant.

Explain maneuverability margins/envelope.

Maneuverability margin/envelope is contained by its upper and lower speed limits, which are either (1) between the aircraft's stall speed (V_S) at the bottom end of its speed range and its VDF/MDF speed at the top end of its speed range or (2) between 1.2/3 V_S (representing a safe operating limit above the stall) at the bottom end of its speed range and VMO/MMO at the top end of its speed range.

What is coffin corner?

Coffin corner occurs at an aircraft's absolute ceiling, where the speeds at which Mach number buffet and prestall buffet occur are coincident, and although trained for, in practice, they are difficult to distinguish between. Therefore, a margin is imposed between an aircraft's operating and absolute ceiling. (*See Q: What is the absolute/maximum service ceiling? page 209.*)

Mach number and the slow-speed stall buffet are coincident at coffin corner because a stall is a function of indicated air speed (IAS) and Mach number is a function of the local speed of sound (LSS), which itself is a function of temperature.

For a constant Mach number (which is the normal mode of speed management), the IAS decreases with altitude due to the decreasing LSS. To prevent the IAS from decreasing to its stall speed, the Mach number must be increased, which results in an increasing IAS.

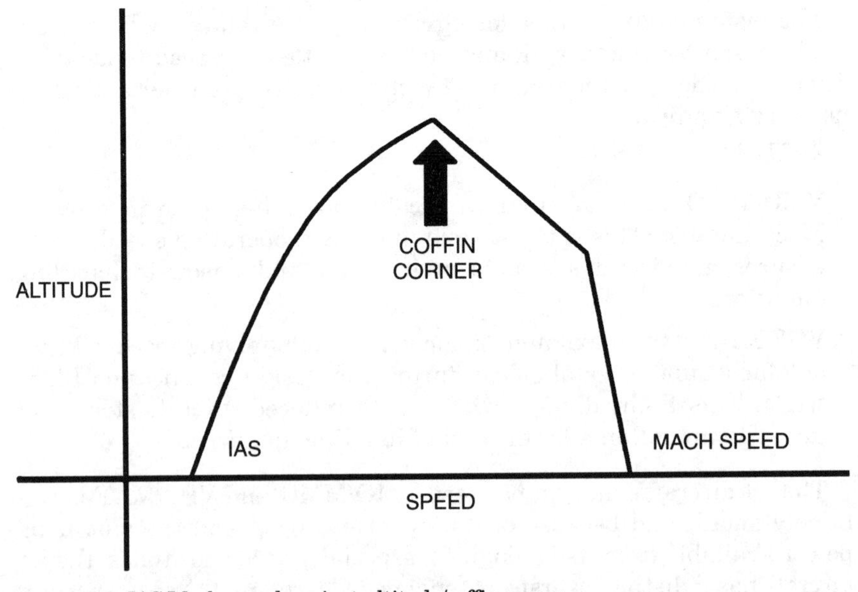

Figure 1.13 IAS/Mach speed against altitude/coffin corner.

For a constant IAS, the Mach number increases with altitude due to a decreasing LSS and temperature to a point where the Mach number exceeds M_{crit}. To prevent the Mach number from exceeding M_{crit}, the IAS must be reduced, which results in a decreasing Mach number.

Therefore, there comes a point at the aircraft's absolute ceiling where the aircraft can go no higher. This is so because it is bounded on one side by the low-speed buffet and on the other by the high-speed buffet because the stall IAS and the M_{crit} values are equal. This is coffin corner, and this effect restricts the altitude attainable by the aircraft.

Stalls

Explain why an aircraft stalls.

An aircraft stalls when the streamlined/laminar airflow (or boundary layer) over the wing's upper surface, which produces lift, breaks away from the surface when the critical angle of attack is exceeded, irrespective of airspeed, and becomes turbulent, causing a loss in lift (i.e., the turbulent air on the upper surface creates a higher air pressure than on the lower surface). The only way to recover is to decrease the angle of attack (i.e., relax the backpressure and/or move the control column forward).

What properties affect an aircraft's stall speed?

An aircraft will stall at a constant angle of attack (known as the *critical angle of attack*). Because most aircraft do not have angle of attack indicators (except "eyebrows" on some electronic flight instrument system displays), the pilot has to rely on airspeed indications. However, the speed at which the aircraft will stall is variable depending on the effects of the following properties.

1. Weight
   a. Actual weight
   b. Load factor, *g* in a turn
   c. Effective weight/center of gravity position
2. Altitude
3. Wing design/lift
4. Configuration
5. Propeller engine power

How does the stall speed vary with weight?

The heavier the aircraft, the higher is the indicated speed at which the aircraft will stall.

If an aircraft's actual weight is increased, the wing must produce more lift (remember that the lift force must equal the weight force), but because the stall occurs at a constant angle of attack, we can only increase lift by increasing speed. Therefore, the stall speed will increase with an increase in the aircraft's actual or effective weight.

The stall speed is proportional to the square root of the aircraft's weight.

What wing design areas delay the breakup of airflow (stall)?

1. Wing slots are the main design feature that delays/suppresses stall speed. A slot is a form of boundary layer control that reenergizes the airflow to delay it over the wing from separating at the normal stall speed. The wing therefore produces a higher coefficient of lift (C_L) and can achieve a lower speed at the stall angle of attack.
2. Lower angle of incidence and a greater chamber for a particular wing section, e.g., wing tips.

What changes the aircraft's angle of attack at the stall?

The movement of the center of pressure point at the stall causes a change in the aircraft's angle of attack.

Normally, a simple swept or tapered wing is designed so that the center of pressure will move rearward at the stall. This is so because the stall normally is induced at the wing root first, where the center of pressure is at its furthest forward point across the wing span. Therefore, the lift produced from the unstalled part of the wing, toward the tips and therefore aft, is behind the root with an overall net result of the center of pressure moving rearward, which results in a stable nose-down change in the aircraft's angle of attack at the stall.

What happens to the stall speeds at very high altitudes, and why?

The stall speed increases at very high altitudes, which the jet aircraft is capable of, because of

1. *Mach number compressibility effect on the wing.* At very high altitudes, the actual equivalent airspeed (EAS) stall speed increases because the Mach number compressibility effect on the wing disturbs the pressure pattern and increases the effective weight on the wing, resulting in a higher EAS stall speed.
2. *Compressibility error on the IAS / ASI(R).* The compressibility correction that forms part of the difference between the indicated airspeed (IAS) and airspeed indicator (reading) [ASI(R)] (which is uncorrected)

and equivalent airspeed (EAS, which is IAS corrected for compressibility and position instrument error) is larger in the EAS to IAS/ASIR direction due to the effect of the Mach number, resulting in a higher IAS stall speed.

How does stall speed vary with center of gravity position?

(*See Q: How does a forward center of gravity position affect the stall speed, and why? page 16.*)

What is a superstall?

A superstall also may be referred to as a *deep stall* or a *Locked in stall condition,* which, as the name suggests, is a stall from which the aircraft is unable to recover. It is associated with rear-engined, high-T-tail, swept-wing aircraft, which because of their design tend to suffer from an increasing nose-up pitch attitude at the stall with an ineffective recovery pitching capability. The BEA Trident crash in 1972 at Slough, England, is probably the most famous and tragic outcome of a superstall.

A superstall has two distinct characteristics:

1. A nose-up pitching tendency
2. An ineffective tailplane

1. The nose-up pitching tendency at the stall is due to
   a. Near the stall speed, the normal rooftop pressure distribution over the wing chord line changes to an increasing-leading-edge peaky pattern because of the enormous suction developed by the nose profile. At the stall, this peak will collapse.
   b. A simple virgin swept- or tapered-wing aircraft will stall at the wing tip first (if the wing has not been designed with any inboard stall properties) mainly due to the greater loading experienced, leading to a higher angle of incidence that causes the wing tip to stall. Because of the wing sweep, the center of pressure moves inboard to a point where it is forward of the center of gravity, therefore creating an increasing pitch-up tendency.
   c. The forward fuselage creates lift, which usually continues to increase with incidence until well past the stall. This destabilizing effect has a significant contribution to the nose-up pitching tendency of the aircraft.

   However, these phenomena themselves are not exclusive to high-T-tail, rear-engined aircraft and alone do not create a superstall. For a superstall to occur, the aircraft will have to be incapable of recovering from the pitch-up tendency at the stall.

2. An ineffective tailplane makes the aircraft incapable of recovering from the stall condition, which is due to
   a. The tailplane being ineffective because the wing wake, which has now become low-energy disturbed/turbulent air, passes aft and immerses the high-set tail when the aircraft stalls. This greatly reduces the tailplane's effectiveness, and thus it loses its pitching capability in the stall, which it requires to recover the aircraft. This is so because a control surface, especially the elevator, requires clean, stable, laminar airflow (high-energy airflow) to be aerodynamically effective.

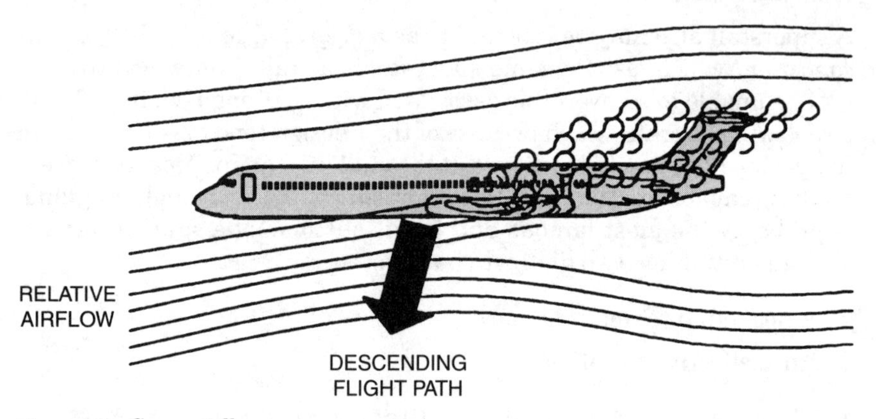

Figure 1.14 Super stall.

What systems protect against a stall?

Stall warner's and stick pushers.

Stall warners are either an artificial audio warning and/or a stick shaker, which usually are activated at or just before the onset of the prestall buffet. Stick pushers are normally used only on aircraft with superstall qualities and usually activate after the stall warning but before the stall, giving an automatic nose-down command. Both systems normally receive a signal from an incidence-measuring probe.

Instability

What is Dutch roll?

Dutch roll is an oscillatory instability associated with swept-wing jet aircraft.

It is the combination of yawing and rolling motions. When the aircraft yaws, it will develop into a roll. The yaw itself is not too significant, but the roll is much more noticeable and unstable. This is so because the aircraft suffers from a continuous reversing rolling action.

What causes Dutch roll?

Swept wings.

Dutch roll occurs when a yaw is induced either by a natural disturbance or by a commanded or an uncommanded yaw input on a swept-wing aircraft. This causes the outer wing to travel faster and to become more straight on to the relative airflow (in effect, decreasing the sweep angle of the wing and increasing its aspect ratio). Both these phenomena will create more lift. At the same time, the inner wing will travel slower and, in effect, becomes more swept relative to the airflow, and both these phenomena will reduce its lift. Therefore, a marked bank occurs to the point where the outer, upward-moving wing stalls and loses all lift, and therefore the wing drops, causing a yaw to the stalled wing and thus leading to the sequence being repeated in the opposite direction. This sequence will continue and produce the oscillatory instability around the longitudinal axis we know as Dutch roll. Pitch fluctuations only occur with an extreme degree of Dutch roll.

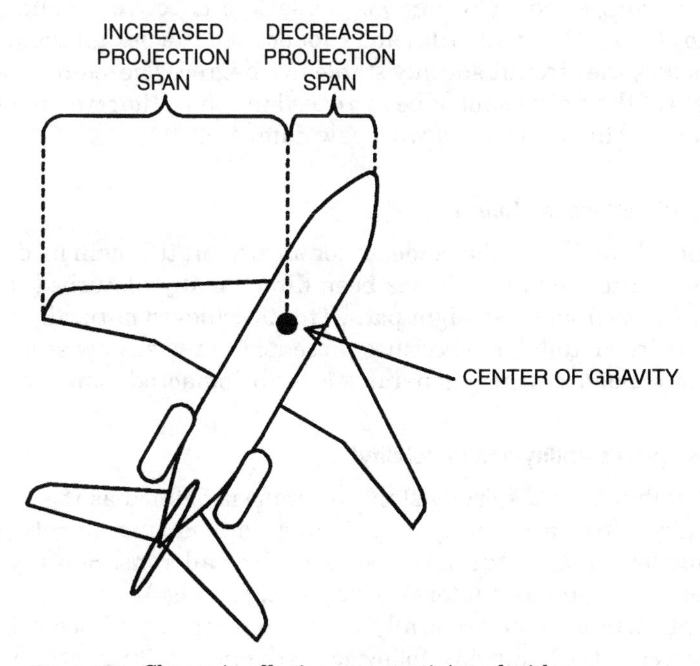

Figure 1.15 Change in effective aspect ratio/speed with yaw.

What is the recovery technique from Dutch roll?

For a pilot to recover an aircraft suffering from Dutch roll, he or she would apply opposite aileron to the direction of the roll, assuming that the yaw dampers are not serviceable.

Although the root cause of Dutch roll is the yawing motion, application of a correcting rudder input by the pilot normally would worsen the situation. This is so because the yawing motion in the oscillatory cycle happens extremely quickly, and the pilot's reaction would not be quick enough to catch the yaw, which already has developed into a roll and dissipated.

Therefore, a rudder input to correct the initial yaw (which has since dissipated) would in fact aggravate the roll effect further into a sideslip.

Aileron control therefore is employed because the roll cycle is of sufficient duration to allow the pilot to apply the correct opposite aileron control. A severe Dutch roll may require two or three aileron inputs to dampen the oscillation gradually.

What prevents Dutch roll?

Yaw dampers prevent Dutch roll on swept-wing aircraft.

A basic reason for the Dutch-rolling tendency of an aircraft (apart from the wing sweep, of course) is the lack of effective fin and rudder area to stop it. The smaller fin and rudder area is a design compromise that makes the aircraft spirally stable to a degree. Therefore, the effectiveness of the fin area must be increased in some other way to prevent Dutch roll. This is achieved with yaw dampers.

What is directional stability?

Directional stability is the tendency for an aircraft to regain its direction (heading) after the aircraft has been directionally disturbed (e.g., an induced yaw) from its straight path. This is achieved naturally because the fin (vertical tailplane) becomes presented to the airflow at a greater angle of incidence, which generates a restoring aerodynamic force.

What is spiral stability and instability?

Spiral stability (or a spirally stable aircraft) is defined as the tendency of an aircraft in a properly coordinated banked turn to return to a laterally level flight attitude on release of the ailerons. Spirally stable aircraft have dominant lateral surfaces (e.g., wings).

Spiral instability or a spirally unstable aircraft will see a banked turn increase fairly quickly, followed by the nose falling into the turn, leading to the aircraft entering into a spiral dive when the ailerons are released in a coordinated turn. Spirally unstable aircraft have dominant (too large) vertical surfaces (e.g., tailplane). What happens is that as the aircraft starts to slip into the turn on release of the ailerons and before the rolling moment due to the sideslip can take effect, the rather dominant fin jumps into play. This is so because the fin/tailplane area (outside) becomes exposed to the relative airflow, which exerts two forces on the aircraft:

1. Around the vertical axis, which straightens the aircraft directionally
2. Around the longitudinal axis, which increases the bank

This accelerates the outer (upper) wing and causes the bank to be increased further. The increased bank causes another slip, which the fin again straightens. This sequence repeats, and the turn is thus made steeper. Once the bank angle exceeds a given type-specific amount (say, 30°), the nose falls into the turn, the speed increases as the roll increases, and the aircraft enters into a spiral dive.

What is lateral stability?

Lateral stability is the tendency for an aircraft to return to a laterally level position around the longitudinal axis on release of the ailerons in a sideslip.

There are two principal features that make an aircraft naturally laterally stable, namely,

1. *Wing dihedral.* The airflow due to a sideslip causes an increase in the angle of attack (lift) on the lower (leading) wing and a decrease in angle of attack on the raised wing because of the dihedral angle. The lower wing thus produces an increase in lift because of the increased angle of attack, and the raised wing produces less lift. The difference in lift causes a rolling moment that tends to restore the wing to its laterally level position.
2. *Side loads produced on the keel surface.* When the aircraft is sideslipping, a side load will be produced on the keel surface, particularly the fin. This side load will produce a moment to roll the aircraft laterally level, which in general terms is stabilizing. The magnitude of this effect depends on the size of the fin, but regardless, its effect is small compared with other laterally stabilizing effects.

What is longitudinal stability?

Longitudinal stability is an aircraft's natural ability to return to a stable pitch position around its lateral axis after a disturbance.

When an aircraft is in equilibrium, the tailplane in general will be producing an up or down load to balance the moments about the center of gravity. (It is assumed that throughout the elevator remains in its original position during any disturbance in pitch.) If the aircraft is disturbed in pitch (say, nose-up), there will be a temporary increase in the angle of attack. The increase in tailplane angle of attack produces an increase in tailplane lift, which will cause a nose-down pitching moment. (The tailplane is thus able to produce a stabilizing moment due to a displacement in pitch as long as the center of gravity remains within its limits.) The wings also experience this increase in angle of attack, resulting in the wings producing an increase in lift. The moment and the direction of the moment produced by this lift will depend on the relative positions of the center of pressure and the center of gravity.

Describe stability at high altitudes.

Longitudinal, lateral, directional, and oscillatory stability in general are reduced at high altitudes, in terms of dynamic stability, mainly because aerodynamic damping decreases with altitude.

The aircraft will feel and is less stable except for spiral stability, which improves with altitude, whereas oscillatory stability deteriorates very rapidly with altitude. This is so because for a constant indicated airspeed (IAS), the fin suffers a smaller angle of incidence and therefore has a smaller restoring force the higher the altitude. Therefore, the fin is less dominant, which is detrimental to oscillatory stability but as a consequence means that the aircraft's lateral surfaces (wings) become more dominant. This improves the aircraft's spiral stability qualities (spiral stability always opposes oscillatory stability, and vice versa).

What are the four flying-quality penalties experienced at very high altitudes?

1. Restricted operating speed range (*See Qs: Explain speed margins? page 25; What is coffin corner? page 26.*)
2. Reduced maneuverability (*See Q: Explain maneuverability margin/envelope, page 26.*)
3. Reduced aerodynamic damping
4. Reduced stability (*See Q: Describe stability at high altitudes, page 34.*)

Control Surfaces

What are the primary/main flight controls?

Elevator. Controls the motion around the lateral axis, known as *pitch/pitching*. (*See Q: What is the elevator, and how does it work? page 35.*)

Ailerons. Control the motion around the longitudinal axis, known as *roll/rolling*. (*See Q: What are ailerons, and how do they work? page 35.*)

Rudder. Controls the motion about the normal/vertical axis, known as *yaw/yawing*. (*See Q: What is the rudder, and how does it work? page 35.*)

What is the elevator, and how does it work?

The conventional elevator is a hinged control surface at the rear of the horizontal tailplane (stabilizer) that is controlled by the pilot's control column.

As the elevator control surface is deflected, the airflow and thus the aerodynamic force around the elevator (horizontal tailplane) changes.

Moving the control column back deflects the elevator up, causing an increase in the airflow speed and thus reducing the static pressure on the underside of the elevator control surface. In addition, the topside of the elevator faces more into the relative airflow, which causes an increase in the dynamic pressure experienced. These effects create an aerodynamic force on the elevator (horizontal tailplane) that rotates (pitches) the aircraft about its lateral axis. That is, back control column movement moves the elevator control surface upward, producing a downward aerodynamic force that pitches the aircraft *up*. Thus the opposite is also true: Forward control column movement moves the elevator control surface downward, producing an upward aerodynamic force that pitches the aircraft *down*.

What are ailerons, and how do they work?

Ailerons are control surfaces located at the trailing edges of the wings that control the aircraft's motion around its longitudinal axis, known as *roll*. The ailerons are controlled by left and right movement of the control column, which commands the ailerons in the following manner: Moving the control column to the left commands the left aileron to be raised, which reduces the lift on the wing, and the right aileron is lowered, which increases the lift generated by this wing, thereby rolling the aircraft into a banked condition, which causes a horizontal lift force (centripetal force) that turns the aircraft. The ailerons normally are powered (hydraulically) powered on heavy/fast aircraft because of the heavy operating forces experienced at high speeds. (*See Q: What are spoilers, and how do they work? page 37.*)

What is the rudder and how does it work?

The rudder is a hinged control surface at the rear of the fin (vertical tailplane) that is controlled by the pilot's rudder pedals.

As the rudder control surface is deflected, the airflow and thus the aerodynamic force around the rudder (vertical tailplane) changes. Moving the left rudder pedal deflects the rudder to the left, causing an increase in the airflow speed and thus reducing the static pressure on the right-hand side of the rudder control surface. In addition, the left side of the rudder faces more into the relative airflow, which causes an increase in the dynamic pressure experienced. These effects create an aerodynamic force to the right on the rudder (vertical tailplane) that rotates (yaws) the aircraft about its vertical/normal axis at its center of gravity point to the left.

How does the effectiveness of the control surfaces vary with speed?

The control surfaces become more effective at higher speeds. This results in a requirement for large control movements at low speeds and smaller control movements at high speeds to produce the same control force.

What is elevator reversal?

Elevator reversal occurs at high speeds when the air loads/forces are large enough to cause a twisting moment on the deflected elevator surface to either a neutral or opposite position that results in sudden reversal of the aircraft's pitch attitude.

What is adverse yaw?

Adverse yaw is a yawing motion opposite to the turning/rolling motion of the aircraft.

Adverse yaw is caused by the drag on the down-going aileron being greater than that on the up-going aileron. This imbalance in drag causes the yawing motion around the normal/vertical axis. Since this yaw is adverse (i.e., in a banked turn to the left, the yaw is to the right), it is opposing the turn, which is detrimental to the aircraft's performance.

Adverse (aileron) yaw is corrected in the design by the use of either differential ailerons or Frise ailerons.

What is aileron reversal (adverse), and when is it likely to occur?

Aileron reversal occurs at high speeds when the air loads/forces are large enough that they cause an increase in lift. Because most of this lift is centered on the down-going aileron at the rear of the wing, a nose-down twisting moment will be caused. This will result in a decrease in the incidence of the wing to the extent that the loss of lift due to the twisting cancels the lift gained from the aileron. At this point the aileron causes no rolling moment, and if the wing twisting is exaggerated (which a down-going aileron can do), the rolling motion around the longitudinal axis can be reversed, hence an adverse rolling motion.

What is a yaw-induced adverse rolling motion, and when is it likely to occur?

The rudder inducing the aircraft to yaw one way can cause another form of adverse rolling motion in the opposite direction. This happens at high speeds (above VMO/MMO) because the deflected rudder experiences a sideways force that causes the aircraft to roll in the opposite direction; i.e., right rudder experiences a sideways force from right to left, causing a rolling moment to the left.

What are spoilers, and how do they work?

Spoilers consist of opening panels that extend from the upper surface of the wing and have the effect of spoiling/disturbing the airflow over the wing (drag), thereby reducing the lift.

For roll control:
The spoilers are raised on one wing and not the other, which creates an imbalance of lift values that produces a rolling moment. The spoilers are connected to the normal aileron controls, and they work in tandem with each other for roll control. Spoilers are in fact a more efficient roll-control surface than ailerons. The disadvantage of roll-control spoilers is that they cause an overall loss of lift, which may cause a loss of height and is particularly undesirable when flying close to the ground.

As air speed brakes:
The spoilers are raised symmetrically on both wings to a flight detent position (using the speed brake lever), which causes a large increase in drag that slows down the speed of the aircraft.

Note. Buffet is usually experienced with spoiler (speed brake) deployment.

As ground lift dumpers:
The spoilers are raised systematically on both wings to the ground detent position (greater angle than the flight detent position), which causes a large increase in drag that (1) decreases lift over the wing, causing the aircraft to sink to the ground, and (2) acts as ground speed brakes to slow down the speed of the aircraft.

What are the three purposes of spoilers?

1. Roll control (usually in combination with the ailerons). Note that the primary purpose of spoilers is roll control.

2. Air speed brakes
3. Ground lift dumpers

Describe differential and nondifferential spoilers.

The difference between differential and nondifferential spoilers is in how they provide lateral roll control when already extended as speed brakes.

Nondifferential spoilers. When already partly extended as a speed brake, the spoilers will extend further on one side but will not retract on the other side in response to a roll command. When already fully extended as a speed brake, both sides remain in the extended speed brake position, and therefore, the spoilers do not provide any roll control.

Differential spoilers. When already partly extended as a speed brake, the spoilers will extend further on one side and retract on the other side in response to a roll command. When already fully extended as a speed brake, the spoilers will remain extended on one side and retract on the other side in response to a roll command.

Give six reasons for spoilers.

Four of the reasons why spoilers are needed on aircraft are to provide a degree of roll control. This is so because the ailerons have the following inadequacies:

1. The ailerons are limited in size and therefore effectiveness.
2. On a thin swept wing, ailerons that are too large will experience a high degree of air loading/lift, resulting in the wing twisting at high speeds that can produce aileron reversal (removes aileron roll control), which is very detrimental. (*See Q: What is aileron reversal (adverse), and when is it likely to occur? page 36.*)
3. Ailerons tend to lose effectiveness at high speeds due to the spanwise diagonal airflow across the aileron, which is less effective than a perpendicular airflow.
4. High-speed swept-winged aircraft cause a strong rolling moment with yaw, known as *adverse rolling moment with yaw.* (*See Q: What is yaw-induced adverse rolling moment, and when is it likely to occur? page 37.*)

Other than roll control, the spoilers are needed to counteract (brake) the aircraft's high speed in the air and on the ground:

5. Because the aircraft has low drag and the engines have a slow lag response rate, there is a need for high-drag devices in flight to act as a brake when the aircraft is required to lose speed and/or height

quickly. This is achieved by the use of the spoilers on both wings being raised simultaneously to the flight detent position, which creates a drag force opposing thrust and therefore reduces the aircraft's speed and/or height.

6. On landing or during a rejected takeoff, there is a need to dump the lift off the wing and onto the wheels to assist in stopping the aircraft. This is achieved by the use of the spoilers on both wings being raised simultaneously to the ground or up detent position in a similar manner as the inflight speed brake. This position has a greater angle of deployment than the flight detent and/or uses more spoiler panels, therefore creating a greater drag force.

What limits the use of spoilers, and why do spoilers blow back?

Spoilers are limited by very high speeds (VDF/MDF), which cause them to blow back.

At very high speeds, the spoilers will be blown back to or near to their fully retracted position. This occurs because the high air loads experienced on the spoilers' surfaces at high speeds are greater than their design limit. Obviously, the force experienced is a function of airspeed and angle of deflection.

How is spoiler blowback prevented?

Spoilers are designed not to blow back in the normal operating speed range of the aircraft. Therefore, correct speed management of the aircraft will prevent the spoilers from blowing back.

How do you correct for spoiler blowback?

In flight, reduce speed by reducing thrust to a speed where the spoilers will operate normally, and then recycle the speed brake lever.

Note: Spoiler blowback will only occur when the aircraft's speed is excessive (i.e., VDF/MDF), which itself should be experienced only in a nonnormal flight condition, e.g., spiral dive, etc., when the recovery drill incorporates reducing speed by closing the throttles.

What do leading-edge slats do?

Leading-edge slats increase the wing's chamber area and mean aerodynamic chord (MAC), thereby increasing its coefficient of lift (C_L) maximum, which reduces the aircraft's stall speed.

What are Krueger flaps?

Krueger flaps are leading-edge wing flaps used to increase the wing chamber and therefore increase the coefficient of lift maximum.

What are Fowler flaps?

Fowler flaps are trailing-edge wing flaps (usually triple slotted) used to increase the wing area and chamber, which increases the coefficient of lift maximum for low flap settings, e.g., 1 to 25°. High flap settings increase drag predominately more than lift and therefore are used to lose speed and/or height, most commonly during an approach to land.

What is the primary use of flaps on a jet aircraft?

The primary use of flaps, especially on a jet aircraft, is to increase lift by extending the geometric chord line of the wing, which increase its chamber and area.

What are the effects of extending flaps in flight?

Lowering the flaps in flight generally will cause a change in the pitching moment. The direction and degree of the change in pitch depend on the relative original position of the center of pressure and the center of gravity.

The factors that contribute to this are

1. The increase in lift created by the increased wing area and chamber will lead to a pitch-up moment if the center of pressure remains in front of the center of gravity.
2. If the associated rearward movement of the center of pressure is behind the center of gravity, then this will produce a nose-down pitch.
3. The flaps will cause an increase in the downwash, and this will reduce the angle of attack of the tailplane, giving a nose-up moment.
4. The increase in drag caused by the flaps will cause a nose-up or nose-down moment depending on whether the flaps are above or below the lateral axis.

The overall change and direction in the pitching moment will depend on which of these effects is predominant. Normally, the increased lift created by extending the wing chord line when the flaps are extended is dominant and will cause a nose-up pitching tendency because the center of pressure normally remains in front of the center of gravity.

What are the effects of raising flaps in flight?

The raising of flaps in flight, if not compensated for by increasing speed and changing attitude, will result in a loss of lift.

How do flaps affect takeoff ground run?

Flaps set within the takeoff range. A higher flap setting, within the takeoff range, will reduce the takeoff ground run for a given aircraft weight. The use of flaps increases the maximum coefficient of lift of the wing due to the increased chord line for a low drag penalty, which reduces the stall speed (V_S) and consequently the rotation (V_R) and takeoff safety (V_2) speeds. This provides good acceleration until it has sufficient kinetic energy to reduce the takeoff ground run. Typically, various flap settings from the first to the penultimate flap setting are available for takeoff (i.e., takeoff range). The higher the flap setting within this range, the less is the takeoff run required because the drag is not significantly increased because the angle of attack is low. However, the drag increment is higher when the aircraft is in flight and out-of-ground effect because of the aircraft's angle of attack is much higher.

Note: Initial and second-segment climb performance thus will be reduced with a high takeoff flap setting. (*See Q: How does the use of flaps affect the aircraft's takeoff performance? page 197.*)

Flaps set outside the takeoff range. A high flap setting outside the takeoff range will result in a large drag penalty that will reduce the aircraft's acceleration, and therefore, the takeoff run will be greatly increased before V_R is attained. No or a very low flap setting outside the takeoff range on takeoff will result in a low coefficient of lift produced by the wing for a given speed, and thus a higher unstick (V_R) speed is required to create the required lift for flight. Therefore, an increased takeoff run to attain the higher V_R is required.

What are the purposes of yaw and roll dampers, and how do they work?

The purpose of a yaw damper is to: (1) prevent Dutch roll and (2) coordinate turns. The purpose of a roll dampers is to: (1) damp/remove Dutch roll, (2) provide roll damping in turbulence, and (3) provide spiral stability.

The yaw damper's main purpose is to prevent Dutch roll. When the fin area is insufficient to provide a natural oscillatory stability, the effective fin area must be increased in some other way. This is accomplished on power-operated rudders with the yaw damper. (*See Q: What causes Dutch roll? page 31.*)

A yaw damper is a gyro system that is sensitive to changes in yaw, and it feeds a signal into the rudder, which applies opposite rudder to the yaw before the roll occurs, thus preventing Dutch roll.

The roll damper works through the aileron controls and can be used to

1. Supplement a yaw damper for Dutch roll control by damping out the roll once it has been established (especially common on oscillatory unstable aircraft).

2. Purely for roll damping in turbulence.
3. Control spiral stability. (*See Q: What is spiral instability? page 32.*)

A roll damper normally is associated with providing spiral stability.

Describe parallel yaw dampers.

Parallel yaw dampers (early type) apply rudder control through the same control run as the pilot, and their activity is reflected in the rudder bar activity (moves the rudder pedals). While this provides a visual indication of the yaw dampers' serviceability, it does increase the rudder loads experienced by the pilot. To prevent this making matters worse in the event of an engine out failure on takeoff or a crosswind landing, the damper can be switched off for takeoffs and landings. Since this damper in effect parallels the pilot's actions, it became known as the *parallel yaw damper.*

Describe series yaw dampers.

The series yaw damper is a development of the parallel yaw damper and is found commonly in modern jet aircraft. This system is attached to the rudder control circuit at the back of the aircraft, and as such, it *does not move the pilot's rudder pedals* when it moves the rudder. This means that the rudder foot forces are not increased and therefore allows the series yaw damper to be used for engine-out takeoffs and landings.

What is a stabilizer/variable-incidence tailplane?

A stabilizer/variable-incidence tailplane is an all-moving horizontal tailplane control surface (i.e., not fixed in one position). (*See Q: What is the purpose of a stabilizer? page 42.*)

Normally, an all-moving horizontal tailplane is called a *stabilizer* when it is solely responsible for longitudinal balancing and it has a separate elevator with its own controls and movement range for pitch maneuverability. A stabilizer normally is moved by its own independent stab trim system that can be either a manual or an automatic device. Normally, an all-moving tailplane is called a *variable-incidence tailplane* when it does not have an elevator surface. Therefore, the variable-incidence tailplane provides pitch maneuverability by control column and longitudinal balancing by the trim system.

What is the purpose of a stabilizer?

The purpose of the horizontal stabilizer is to provide a longitudinal balancing force to the aircraft. Thus the elevator range and aircraft's pitch maneuverability are not compromised and remain available to be used solely to control the pitch of the aircraft. The stabilizer covers large and small pitching moments; e.g., a single person who moves from the aft to the forward cabin will displace the overall weight balance.

What are the four reasons for a variable-incidence tailplane/stabilizer, especially on a jet aircraft?

The four reasons for use of a variable-incidence tailplane or stabilizer are

1. To provide a balancing force for a large center of gravity range
2. To provide a balancing force for a large speed range
3. To cope (longitudinally) with large trim changes as a result of position changes to the wing leading and trailing edge high-lift devices (configuration changes)
4. To reduce elevator trim drag to a minimum

While any of these requirements in isolation might not demand a variable-incidence tailplane, in combination they certainly do. Once the need has been established for one of the requirements, then advantages also occur in the other areas.

Describe the effects of a stuck stabilizer.

The condition an aircraft experiences from a stuck stabilizer, other than when the aircraft's equilibrium is matched to the position of the stuck stabilizer, is a degraded longitudinal balancing ability, which is due to the backup employment of the less powerful elevator in providing this balancing force.

This condition has the following effects:

1. The longitudinal balancing force the elevator experiences is much higher than when it is used for its normal pitch control duties, and this is translated into heavy stick forces felt by the pilot.
2. Because the elevator surface is smaller than the stabilizer, the response to the pilot's inputs is slower.
3. The elevator's pitch-control capability is reduced because the majority of its range is being used to provide a longitudinal balancing force.

What is the best center of gravity position with a stuck stabilizer, and why?.

With a stuck stabilizer, the best center of gravity position is *aft*. This can be accomplished by moving passengers to the rear of the aircraft and/or by moving fuel to outer wing tanks if possible.

What is the required action with a stuck stabilizer?

If a stabilizer has stuck while the aircraft is in a substantially trimmed-out condition, then as long as you maintain the speed at which the tail jammed, you will remain substantially in trim and therefore stable. However, because the aircraft will have to depart from the cruise speed in preparation to land, the following steps will reduce the main reasons for having a variable-incidence tailplane in the first place and also reduce the effects of a stuck stabilizer. (*See Q: Describe the effects of a stuck stabilizer? page 43.*) This helps to maintain the aircraft's longitudinal stability and maneuverability in pitch during its approach and landing. Generally, the following procedure would be followed:

1. Divert to a nearby airfield (so that the center of gravity movement due to the aircraft's change in weight is not excessive over a prolonged flight).
2. Move the center of gravity to an aft position. This will
   a. Longitudinally balance the aircraft at low speeds and thereby reduce the stick forces.
   b. Reduce the elevator demand for the landing flare. (*See Q: What is the best center of gravity position with a stuck stabilizer, and why? page 43.*)
3. Reduce speed as late as possible to minimize the length of time a balancing force, with its associated high stick forces, is required from the elevator.
4. Plan a long final approach, and make configuration changes, gear, and flaps earlier than usual to give time to sort out the aircraft before the next change is due.
5. Use a reduced flap setting for landing, which will reduce the landing flare required. This allows you to maintain a higher approach speed, which reduces the divergence of the aircraft from its cruise trimmed speed, reduces the balancing force required from the elevator, and thus reduces the stick forces experienced.

Having exercised these measures in response to this failure, the aircraft will retain enough scope to maintain a longitudinally stable condition and enough elevator pitch maneuverability to adjust the aircraft's approach and landing attitude, which should not produce too many problems for the pilot.

Describe a runaway stabilizer condition and required action.

A stabilizer typically is held in its trimmed position by a series of brakes for both manual and autopilot modes. If these brakes should fail, then the stabilizer will experience backpressure from the airflow, which will rotate the stabilizer to its maximum upward or downward mechanical stops, thereby inducing a marked out-of-trim unstable condition.

A stabilizer runaway, whether auto or manual trim, should not go undetected for too long, and when detected, decisive action should be employed.

Each aircraft type has its own drill, but in general terms:

1. Hold the control column firmly.
2. Autopilot (if engaged), disengage.
3. Stab trim cutout switches to cut out.

If the runaway continues,

4. Stabilizer trim wheel, grasp and hold.

Continue your flight using manual trim, and adopt early airspeed and configuration conditions.

Describe the effects of a jammed/degraded elevator.

A degraded or jammed elevator will result in less effective elevator maneuverability in pitch control.

What is the best center of gravity position with a jammed/degraded elevator?

The best position for the center of gravity with a degraded/jammed elevator is *aft*. This can be accomplished by moving the passengers to the rear of the cabin and/or by moving fuel to the outer wing tanks if possible. An aft center of gravity position lessens the need for large pitch-control demands, especially during the approach and landing flare.

What is the required action for a jammed/degraded elevator?

If there is *no* elevator control (jammed), then the stabilizer trim can be used for pitch control. If the elevator control is reduced (degraded), then it should be assessed if there is still sufficient elevator range available to land safely and used to do so if applicable. If not, then the condition should be regarded as having no elevator control.

With a degraded or jammed elevator, several actions can be taken to minimize the need for major pitch changes and/or to improve the handling and management of the aircraft:

1. If possible, move the center of gravity rearward; this will reduce the need for large elevator angles, especially during the landing.
2. Plan a long final approach, and make configuration changes, gear, and flaps earlier than usual to allow more time to sort out the aircraft before the next change has to be made.
3. Restrict the flap angle for landing to reduce the flare demanded.

Describe the effects of a failure/reduction in elevator feels.

Artificial feel systems (normally duplicated) are employed on powered controls, especially the elevators. They meet the requirement of progressive feel against control surface deflection at constant speed and against a constant angle at varying speed based on our old friend

$$\tfrac{1}{2} R\, V^2$$

Whenever the feel on an elevator control is significantly reduced, great care must be exercised in its use. The control must be moved slowly and smoothly over minimum angles to avoid overstressing the control surface structure but enough to maintain the flight path. Overstressing the control surface with a lack of feel is a significant problem, and for this reason, turbulence should be avoided.

What is the best center of gravity position with a reduced or failed elevator feel system?

The best position for the center of gravity with a reduced or failed elevator feel system is *forward.* This can be accomplished by moving the passengers to the front of the cabin and/or by moving the fuel into forward tanks if possible.

This increases the aircraft's natural longitudinal stability and renders the pitch control less sensitive and feeling heavier, therefore making the aircraft less responsive to small elevator movements. Thus the chances of the pilot overstressing the elevator control surface are minimized, although still possible.

Describe the effects of the air loads on a control surface and how these effects are managed?

If a control surface is deflected, the dynamic pressure/aerodynamic loads on it will increase and act as a lift force through its center of pressure point. When multiplied by the control surface arm, it gives the size of the moment trying to rotate the control surface back to its neutral position. This moment is known as the *hinge moment* or *air load force.* That is,

$$\text{Hinge moment (air loads force)} = \text{lift force (air load)} \times \text{arm}$$

Note: Where the lift force is a design product of the size of the control surface and the magnitude of the lift force experienced depends on (1) airspeed and (2) angle of deflection of the control surface. That is, the lift force increases dynamically in flight with either an increased control angle of deflection at a constant speed or at a constant angle of deflection with an increased speed. And where the arm is a design product of the distance between the center of pressure point and the hinge line. They produce a hinge moment/air load force that tries to return the surface to its neutral position.

What is a tab surface, and what can it be used for?

A tab is a small hinged surface found on a primary flight control surface. Usually the tab forms part of the trailing edge on a manual control surface. A tab can be used to provide

1. Trimming
2. Control balance
3. Servo operation of a control surface

to reduce/balance the opposing hinge moment (air load force) on the associated control surface.

What is a balance tab?

A balance tab is a form of aerodynamic control balance on a control surface. A control balance tab balances the main aerodynamic lift force load on a control surface with an opposing force, which thereby reduces the overall hinge moment (air load force).

$$\text{Hinge moment (air loads)} = \text{lift force} \times \text{arm}$$

This is reflected by the stick control force that the pilot experiences being reduced to a manageable level. (*See Q: Describe the effects of the air loads on a control surface and how these effects are managed, page 46.*)

What is a hinge/horn balance?

A setback hinge is another form of aerodynamic control balance on a control surface, whereby the design of the control surface sets the hinge line back into the control surface, thus reducing the center of pressure to hinge line arm, which results in reducing the control surface hinge moment. [Remember, hinge moment (air loads) = lift force × arm.] Thus the stick control force that reflects the overall hinge moment experienced on the control surface is reduced to a manageable level for the pilot. (*See Q: Describe the effects of the air loads on a control surface and how these effects are managed, page 46.*)

Another form of aerodynamic control balance on a control surface is a horn balance. A horn balance is a protruding control surface that produces a balancing lift force in the opposite direction of the main lift force and reduces the overall hinge moment/air load force. They are common on elevator/stabilizers. Thus the stick control force that reflects the overall hinge moment experienced on the control surface is reduced to a manageable level for the pilot.

Note: A horn balance and setback hinge typically are used in tandem on a control surface.

What is a mass balance and what is it used for?

A mass balance is another form of aerodynamic balance control on a control surface. The hinge moment/air load force experienced by a deflected control surface tries to rotate the control surface back to its neutral position, but it is balanced by a mass weight that keeps it in its deflected position. Thus the stick control force that reflects the overall air load force/hinge moment experienced on the control surface is reduced to a manageable level.

Why are control surfaces hydraulically operated on large aircraft?

On large, fast aircraft, especially modern jets, it is found that the control forces required to move a control surface are simply beyond the strength of the pilot and are also too great to be controlled by pure aerodynamic designs, e.g., balance tabs. This is so because the shear sizes and weights of the control surface arms in question and the aerodynamic airflow lift forces (load) generated on the deflected control surface are too great.

For modern large, fast jet aircraft, the answer lies in the powered control surface, typically hydraulic-powered systems, because they generate enough power to cope with the full air load force (i.e., not balanced) experienced on the control surface.

Why does a powered controlled surface need an artificial feel system?

An artificial feel system is required because power-operated flying controls are irreversible, i.e., they do not feed back to the pilot any sensory information about how hard the control surface is and thus what aerodynamic air forces it is coping with. Therefore, there is a need to give this information to the pilot so that he or she is aware of the control angles being applied and their effect on the aircraft, in short, keeping the pilot in the sensory loop, which allows him or her to guard against overstressing the control surface.

How does an artificial feel system work?

The simplest form of artificial feel consists of a spring box fitted into the control run. This provides a feel and self-centering action, but the stick forces are constant and therefore are only suitable for aircraft types with a limited altitude and speed range.

What is Q feel, and where is it used?

Q feel is a sophisticated computer-based artificial feel system based on $\frac{1}{2} R V^2$ that is felt by the pilot through the control column and rudder pedals and is used commonly on aircraft with powered flight controls, i.e., elevator, rudder, and ailerons. It meets the requirements of progressive feel to match variable control surface deflection at a constant speed and/or for a constant angle of deflection at varying speeds. (*See Q: How does an artificial feel system work? page 48.*)

What are the inputs to Q feel?

1. Static and dynamic pressure
2. Control surface angle of deflection

What are active controls?

An active control is a surface that moves automatically/actively in response to nondirect inputs. For example, balance tabs actively/automatically move in response to their associated control surfaces being moved. Auto slats actively/automatically move to their full extend position to provide a better coefficient of lift (C_L) on a B737-300 if the aircraft senses a particular flight condition; i.e., trailing-edge flaps set 1 to 5 positions, auto slats in normal position, and aircraft close to the stall.

Chapter

2

Engines

Applied Laws

What is thermodynamics?

Thermodynamics is the study of heat/pressure energy or the behavior of gases (including air) and vapors under variations of temperature and pressure.

Explain Bernoulli's theorem.

Bernoulli's theorem is that the total energy in a moving fluid or gas is made up of three forms of energy:

1. Potential energy (the energy due to the position)
2. Pressure/temperature energy (the energy due to the pressure)
3. Kinetic energy (the energy due to the movement)

When considering the flow of air, the potential energy can be ignored; therefore, for practical purposes, it can be said that the kinetic energy plus the pressure/temperature energy of a smooth flow of air is always constant. Thus, if the kinetic energy is increased, the pressure/temperature energy drops proportionally, and vice versa, so as to keep the total energy constant. This is Bernoulli's theorem.

Explain a venturi.

A venturi is a practical application of Bernoulli's theorem, sometimes called a *convergent/divergent duct.*

A venturi tube has an inlet that narrows to a throat, forming a converging duct and resulting in (1) velocity increasing, pressure (static) decreasing, and (3) temperature decreasing. The outlet section is relatively longer with an increasing diameter, forming a diverging duct and resulting in (1) velocity decreasing, (2) pressure (static) increasing, and (3) temperature increasing.

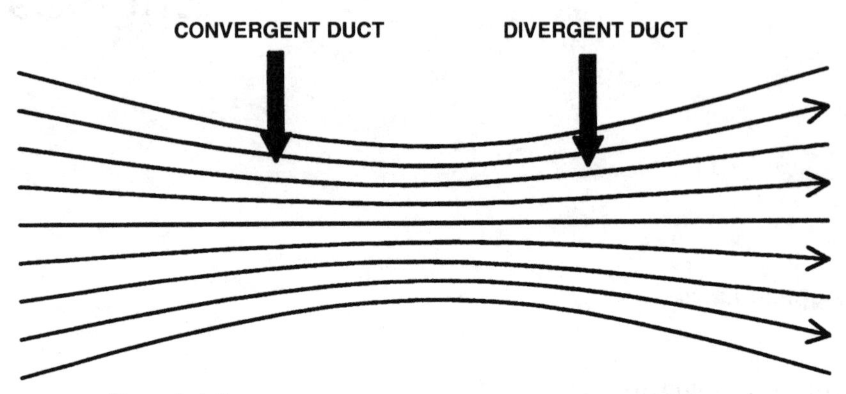

Figure 2.1 Venturi airflow.

For a flow of air to remain streamlined, the mass flow through a venturi must remain constant. To do this and still pass through the reduced cross section of the venturi throat, the speed of flow through the throat must be increased. In accordance with Bernoulli's theorem, this brings about an accompanying drop in pressure and temperature. As the venturi becomes a divergent duct, the speed reduces, and thus the pressure and temperature increase.

Piston Engines

What is the combustion cycle of an aeropiston engine?

Induction, compression, combustion (expansion), and exhaust. The combustion of a piston engine occurs at a constant volume.

What is compression ratio in a piston engine?

Compression ratio in a piston engine is the ratio of the total volume enclosed in a cylinder with the piston at bottom dead center (BDC) to the volume remaining at the end of the compression stroke with the piston at top dead center (TDC).

$$\text{Compression ratio} = \frac{\text{total volume}}{\text{clearance volume}}$$

What produces the ignition in a piston engine?

Magnetos are used on internal combustion engines to supply the high-tension voltage necessary to cause an electric spark at the spark plug.

What does blue, black, and white exhaust smoke indicate?

Blue exhaust smoke indicates an oil burn in the cylinders, probably due to broken piston rings that allow oil seepage into the combustion chamber. Black exhaust smoke indicates carbon granules burning in the cylinders. This occurs if the mixture is too rich, resulting in some of the fuel not being burnt and turning into carbon granules, which are then exhausted as black smoke. White exhaust smoke indicates a high water content in the combustion chamber, which is exhausted as white "steam" smoke.

What is engine torque?

Torque is a force causing rotation. In particular, torque is the force created within an engine, which causes rotation of the rotating parts, e.g., the crankshaft.

Torque is a measure of the load experience, expressed in pounds per inch or feet. A quantity of torque, or twisting moment, is involved in the measurement of the engine brake horsepower (bhp).

$$\text{Torque} = \text{force} \times \text{distance} \qquad \text{(at right angles to the force)}$$

What is a normally aspirated piston engine?

A normally aspirated engine works as a result of "the breathing of the cylinder is due to the pressure differential below the standard sea level 14.7 lb/in^2." In other words, it only uses the atmospheric air density that is available to produce a charge in its cylinders and is not boosted by a supercharger; therefore, the normally aspirated piston engine's power output is restricted by its cylinder capacity.

What are the disadvantages of a piston engine?

A piston engine suffers from three main disadvantages:

1. A lack of power output, especially with increased altitudes
2. A low produced airspeed due to propeller rpm limitations
3. Mechanical inefficiency

What is a supercharged (piston) engine?

A supercharger simply increases the air delivered to the engine cylinder above its normally aspirated capacity by compressing the intake air, which in turn requires more fuel to be delivered to the carburetor to maintain the correct mixture ratio, which in turn produces a greater power output (horsepower). Therefore, a supercharged engine is capable of producing a greater power output than a normally aspirated engine of the same cylinder size.

How is the (piston) engine power output increased to compensate for low atmospheric pressure?

Superchargers are used to artificially raise the engine manifold pressure to compensate for low atmospheric pressure either (1) to increase engine power output for takeoff and the initial climb (ground-boosted engine) or (2) to maintain mean sea level (MSL) engine power at high altitudes (altitude-boosted engine).

What regulates the supercharger to deliver a constant boost/manifold absolute pressure (MAP)?

The auto boost control (ABC) keeps the boost pressure/MAP constant during a climb or a descent.

How is engine power monitored?

There are two main engine power monitoring/indication systems: manifold absolute pressure (MAP) and boost pressure.

What is carburetor icing?

(*See Chapter 8, "Meteorology and Weather Recognition," page 264.*)

When would you expect carburetor icing in a piston engine?

(*See Chapter 9, "Flight Operations and Technique," page 325.*)

What actions should you take to prevent or remove carburetor/throttle icing in a piston engine?

(*See Chapter 9, "Flight Operations and Technique," page 325.*)

Propellers

What advantages does an aircraft gain from a propeller?

The propeller provides the following advantages:

1. The propeller creates a high-energy slipstream, which has three main effects on the aircraft:

   a. The slipstream creates extra lift over the wing.
   b. The slipstream suppresses the stall speed of the aircraft.
   c. The slipstream makes the fin/rudder more effective.
2. The propeller/piston engine has a quick response rate to a throttle input, which gives an early application of the slipstream effects on the aircraft.

Therefore, a propeller-driven aircraft has good slow-speed recovery.

What is the main disadvantage of a propeller?

A propeller-driven aircraft suffers from a lack of airspeed due to propeller rpm limitations as a result of propeller compressibility losses.

This is so because the propeller suffers the effects of compressibility when the speed of the propeller blade tips becomes sonic. Therefore, propellers have a revolution speed limited to just below sonic speed, which limits the thrust force produced by propellers and therefore the aircraft's airspeed. This results in (1) a lower true airspeed (TAS) and (2) a shorter range.

What produces thrust on a propeller-driven aircraft?

The forward thrust (horizontal lift) of a propeller-driven aircraft is a result of Newton's third law. "The application of a force on a body will cause an equal and opposite reaction." The propeller drives an accelerated mass of air rearward, and the thrust force developed by the propeller is equal to the mass of the air and the rate of change in momentum given to the air, which has the reaction of driving the aircraft forward.

The amount of the force created by the propeller is governed by two factors:

1. The output of the engine, which drives the prop shaft.
2. The blade angle, angle of attack, and pitch of the propeller.

What restrictions does the propeller design have?

The degree to which the propeller design can be increased to absorb engine power is restricted in the following areas:

1. Blade length is restricted because of the following:
   a. The need for adequate ground and fuselage clearance.
   b. The need to maintain subsonic blade-tip speeds (the longer the propeller blade, the greater are the tip speeds, and supersonic propeller speeds are inefficient).
2. Blade chord size is restricted because of the following:

   a. An increase in the chord size will reduce the aspect ratio (blade diameter/chord), and as with the wing, this gives a lower efficiency output than the optimal ratio.
   b. Larger chord lengths also increase the centrifugal twisting moment, which tends to twist the blade to a finer pitch, causing
      (1) Higher loads on the root fittings.
      (2) Higher torque values, which give a lower resultant force/lower thrust output (forward thrust force minus torque force equals resultant force).
3. The number of blades can be increased to a certain value, but with more blades, the hub diameter and weight become excessive, and blade interference begins to reduce efficiency.

How does the propeller convert engine horsepower to produce thrust?

The propeller is connected to the engine via a prop shaft that rotates the propeller, which generates an accelerated mass of air rearward, thereby converting the shaft horsepower of the engine into a thrust force.

$$\text{Thrust force} = \text{air mass} \times \text{velocity}$$

Air mass is determined by (1) blade angle, (2) blade angle of attack, and (3) pitch. Air velocity is determined by propeller rpm, which is set by (1) engine power output and (2) blade angle of attack.

Why is the propeller blade twisted?

The propeller blade is twisted along its length to maintain a constant blade angle of attack.

How do you define propeller efficiency?

The propeller efficiency in producing thrust to propel an aircraft forward is determined as the ratio of the useful work done by the propeller (propeller thrust) in moving the aircraft to the work supplied by the engine (engine bhp).

$$\text{Propeller efficiency} = \frac{\text{prop thrust (air mass} \times \text{velocity)}}{\text{engine brake horsepower (bhp)}}$$

What are the disadvantages of a fixed-pitch propeller?

The failing of a fixed-pitch propeller is that it only produces its maximum efficiency (i.e., best blade angle/pitch for rpm speed) at one predetermined engine rpm, altitude, and forward airspeed condition. This is so because the forward airspeed affects the blade angle of attack (an increase in forward airspeed causes a decrease in blade angle of attack, which reduces the rearward air displacement and thereby decreases the thrust produced), and therefore, propeller efficiency is reduced away from its single predetermined condition. (*See Q: How do you define propeller efficiency? page 56.*)

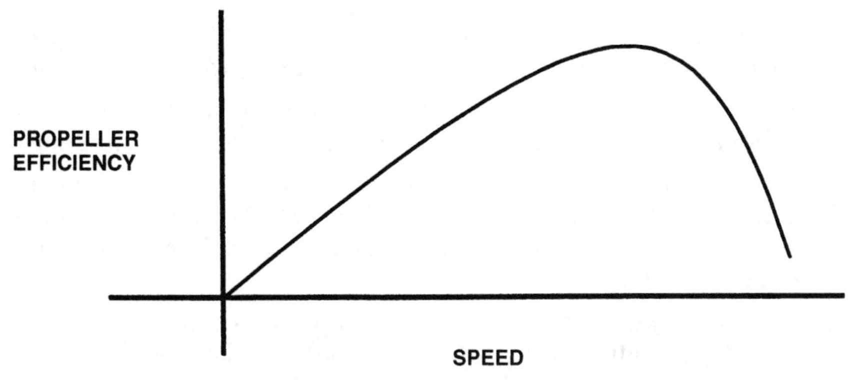

Figure 2.2 Propeller efficiency/blade angle or pitch against speed.

Therefore, for a fixed-pitch propeller, a single setting must be made between the blade angle (and therefore the angle of attack and pitch) desirable for takeoff and that required for cruise conditions. And if a blade angle for high-speed cruise is chosen (coarse), then the aircraft has the disadvantage of poor propeller performance for takeoff, and vice versa. (*See Q: What is a variable-pitch propeller, and why is it used? page 57.*)

What is a variable-pitch propeller, and why is it used?

Variable-pitch propellers are a development of the fixed-pitch propeller (*see Q: What are the disadvantages of a fixed-pitch propeller? page 56*) that have a variable and controllable blade angle between the coarse and fine positions. This is used to adjust the blade angle of attack to its optimal setting in order to maintain propeller efficiency and aircraft thrust over a wide range of aircraft speeds, phases of flight, and differing operating conditions. A variable-pitch propeller maximizes the propeller's efficiency through a large speed range by maintaining a constant blade angle of attack that thereby produces a constant thrust value.

What controls the propeller blade angle/speed?

A constant-speed unit (CSU) (sometimes called a *governor unit*) controls the propeller's blade angle/pitch to maintain an rpm speed.

Why is a turboprop aircraft better suited for short regional operations?

Turboprop aircraft in general are better suited for short regional operations because short sector routes normally have a restricted cruise altitude, which the turboprop aircraft is normally better suited to for the following reasons:

1. A jet engine is suited to high altitudes (i.e., 30,000 ft and above), which an aircraft is not capable of reaching during a short sector route.
2. The turboprop engine is designed to operate at its most efficient at a medium altitude, which is associated with short regional operations.
3. In addition, many short regional operations operate out of restrictive airfields, and the turboprop aircraft's high-lift straight wing is capable of meeting the field lengths and climb and descent gradients of these restrictive airfields.
4. In addition, short-range sectors usually are flown more frequently and with a smaller passenger demand per trip. Therefore, a turboprop aircraft, which typically has a smaller passenger capacity than a jet aircraft, is a better economical design for this type of short sector demand, i.e., 20- to 60-passenger loads.

However, it should be noted that this trend is changing, with 30- to 70-seat jet airliners now being built that are efficient at medium altitudes over short regional routes and as such are replacing the turboprop fleets operated by regional airlines.

Is there a critical engine on a propeller aircraft?

Yes, especially on aircraft with propellers rotating in the same direction. A critical engine is the engine that determines the critical control speed of the aircraft.

Why is the number 1 engine the critical engine on a multiengine propeller aircraft?

There are two main reasons why the number 1 engine is the critical engine on an aircraft with propellers rotating in the same direction: (1) slipstream effect and (2) asymmetric blade effect.

1. *Slipstream effect.* If the propellers are rotating in the same direction (i.e., clockwise viewed from behind and counterclockwise viewed from in front), then only the number 1 engine will produce a sideways slipstream force on the fin. This has the effect of assisting the rudder side force needed to counteract the yawing moment for a number 2 engine failure. However, for a number 1 engine failure, the slipstream from the number 2 engine will produce a sideways force that aggravates the yawing moment, resulting in a more critical situation.

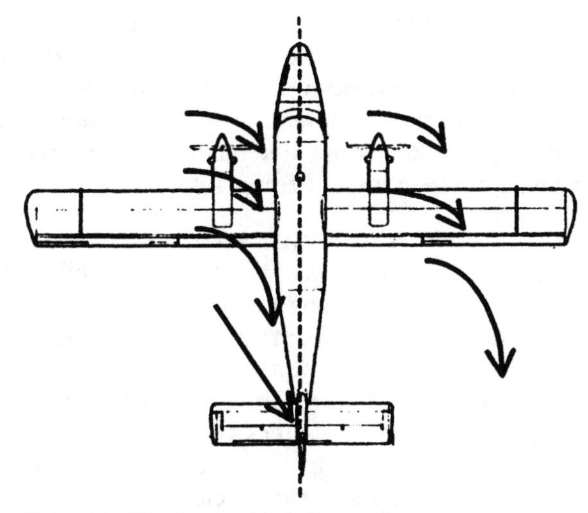

Figure 2.3 Slipstream effect of propellers.

Therefore, a greater control force is required, resulting in a need for a higher critical speed to make the rudder more effective during a number 1 engine failure, which therefore determines that the number 1 engine is the critical engine.

2. *Asymmetric blade effect.* Propeller blades produce more thrust in the downward rotation than in the upward rotation. Therefore, the point through which the thrust acts will be displaced toward the down-going

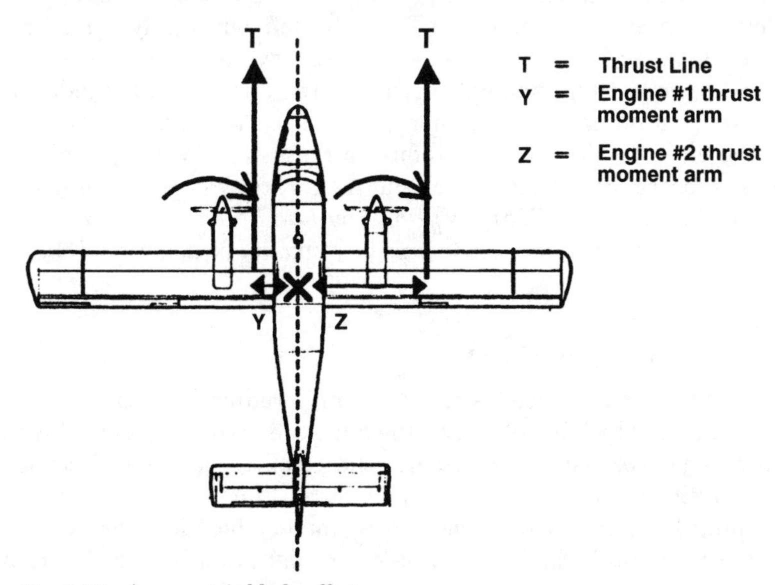

Figure 2.4 Asymmetric blade effects.

blade. Depending on the direction of rotation of the propellers, this either increases or decreases the thrust moment arm, and in aircraft with propellers rotating in the same direction, failure of the engine with the shortest moment arm will produce the greatest yawing moment from the other live engine (i.e., normally counterclockwise rotation when viewed from in front).

Therefore, the engine with the shortest thrust moment arm is the critical engine, and for counterclockwise rotating propellers, this makes the number 1 engine the critical engine.

It can be seen that the aircraft's critical speed is based on the critical engine and is proportional to the length of the thrust moment arm; i.e., the shorter the thrust moment arm, the lower is the critical speed.

The use of contrarotating propellers (i.e., engines with propellers rotating in opposite directions), e.g., number 1 engine rotating counterclockwise and number 2 engine rotating clockwise when viewed from in front, reduces the difficulties of asymmetric blade.

If both propellers rotate inward, toward the fuselage, then (1) the thrust moment arm will be kept as short as possible, and (2) slipstream effects counteract the yawing moment for both engine failures. Consequently, the aircraft's critical speed is reduced.

How does a crosswind affect the critical engine?

A crosswind, depending on its direction, can either help to restore or aggravate the yawing moment of an aircraft with a failed critical engine.

For instance, a failed critical number 1 engine will cause a yaw to the left. A crosswind component from the left will apply a restoring force to the aircraft's fuselage, whereas a crosswind from the right will aggravate the yawing moment further to the left due to the sideways force experienced on the right side of the aircraft's fuselage (which is from the right to the left). Therefore, a crosswind landing is of even greater importance with a critical engine failure. (*See Q: If you had an engine failure between V_1 and V_R and you had a maximum crosswind, which engine would be the best to lose, i.e., upwind or downwind engine? page 195.*)

What is a windmilling propeller?

Windmilling is experienced when the rpm is reduced but the airspeed is maintained, which eventually will cause the blade angle of attack to become negative. When this occurs, the resulting force will act in a rearward direction. This is known as *windmilling*.

A windmilling propeller causes a drag force, which is opposite to the direction of flight. This drag can be quite high, and in addition to its decelerating effect, it also can cause a large yawing moment on a multi-engined aircraft with asymmetric thrust.

Can you obtain ground reverse/braking thrust from propellers?

It is possible to increase the braking effect on some aircraft types by reversing the pitch of the propeller (negative angle), causing it to produce thrust in the opposite direction.

What is propeller feathering, and why is it used?

Feathering the propeller is accomplished in the event of an inflight engine failure or fire.

Feathering a propeller is the term given to the arrangement of the propeller blades turned beyond the coarse pitch until they are edge on (i.e., the chord line is parallel) to the airflow, thereby causing the propeller *not* to windmill (or create extra drag) or hinder the extinguishing of an engine fire.

How do propeller aircraft generate noise?

The noise generated by a propeller aircraft is from the sheer effect of different displaced air velocities. That is, the sheer is the difference between the propeller's faster displaced air and the slower ambient air around it.

How is propeller noise controlled or reduced?

The following can reduce propeller noise:

1. Increase the number of propeller blades, i.e., four blades instead of three. This increases the mass of airflow; therefore, its velocity can be reduced to maintain the same thrust. This results in reducing the sheer effect between the displaced propeller air and the ambient air and reduces noise.
2. Reduced-thrust takeoff.

The following can control propeller noise: maximum-angle climb after takeoff. This allows the aircraft to get above any noise control zones.

Jet/Gas Turbine Engines

What is the theory of a jet/gas turbine engine?

Frank Whittle described the theory behind the jet engine as the balloon theory: "When you let air out of a balloon, a reaction propels the balloon in the opposite direction." This, of course, is a practical application of Newton's third law of motion.

A jet/gas turbine produces thrust in a similar way to the piston engine/propeller combination by propelling the aircraft forward as a result of thrusting a large weight of air rearward.

$$\text{Thrust} = \text{air mass} \times \text{velocity}$$

Early jet engines adopted the principle of taking a small mass of air and expelling it at an extremely high velocity. Later gas turbine engines have evolved into taking and producing a large mass of air and expelling it at a relatively slow velocity (e.g., high-bypass engine).

What is specific fuel consumption (SFC)?

Specific fuel consumption is the quantity/weight (lb) of fuel consumed per hour divided by the thrust of an engine in pounds:

$$\frac{\text{Fuel burn (lb) per hour}}{\text{engine thrust (lb)}}$$

What is the combustion cycle of a jet/gas turbine engine?

The combustion cycle of a jet/gas turbine engine is induction, compression, combustion, expansion, and exhaust. In a jet/gas turbine engine, combustion occurs at a constant pressure, whereas in a piston engine, it occurs at a constant volume.

Why was the jet/gas turbine engine invented?

Frank Whittle invented the jet aircraft engine as a means of increasing an aircraft's attainable altitude, airspeed, reliability, and, to a lesser extent, maneuverability for the military.

Frank Whittle designed the jet/gas turbine engine for two main reasons:

1. To achieve higher altitudes and thus airspeed because propeller aircraft had limited altitude and speed capabilities.
2. As a more simplistic and therefore reliable engine because the piston engine was a very complicated engine with many moving parts and thus was unreliable.

Describe how a jet/gas turbine engine works.

The jet engine or aerothermodynamic duct, to give it its real name, has no major rotating parts and consists of a duct with a divergent entry and convergent or convergent/divergent exit. When forward motion is imparted to it from an external source, air is forced into the engine intake, where it loses velocity or kinetic energy and therefore increases its pressure energy as it passes through the divergent duct. The total energy is then increased by the combustion of fuel, and the expanding gases accelerate to atmosphere through the outlet converging duct, thereby producing a propulsive jet.

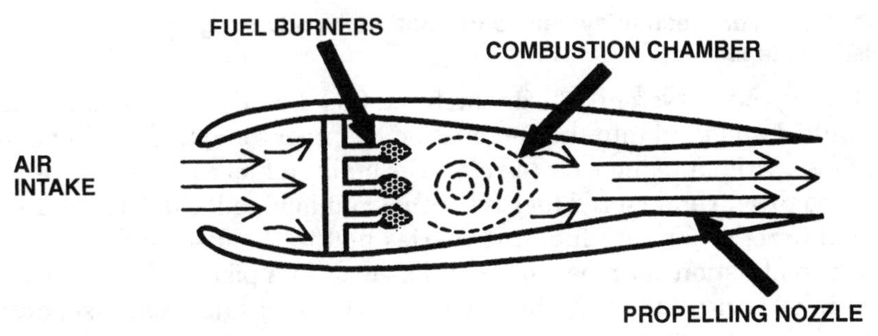

Figure 2.5 The jet engine.

A jet engine is unsuitable as an aircraft power plant because it is incapable of producing thrust at low speeds. That is, it requires a forward motion itself before it produces any thrust.

The gas turbine engine has avoided the inherent weakness of the jet engine by introducing a turbine-driven compressor that produces thrust at low speeds. Therefore, the aircraft power plant is in fact a gas turbine engine and subsequently will be referred to as such.

The gas turbine is essentially a heat engine using air as a working fluid to provide thrust by accelerating air through the engine and increasing its kinetic energy. To obtain this increase, the pressure energy is increased first by a *compressor,* followed by the addition of heat energy in the *combustion chamber,* before its final conversion back to kinetic energy in the form of a high-velocity jet efflux across the *turbine.* (This provides extra shaft power to either drive a conventional frontal propeller or fan or to compress extra air to provide more jet flow, as in a ducted fan and bypass engine.) The airflow is then finally exhausted through the *exhaust* nozzle duct.

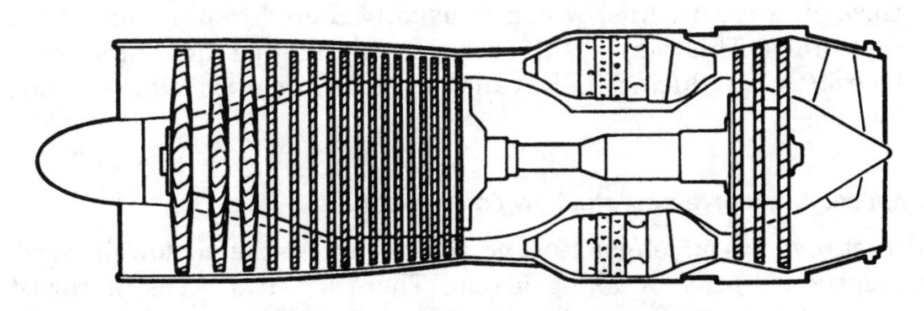

Figure 2.6 Gas turbine engine—single-spool axial-flow compressor. (*Reproduced with kind permission of Rolls-Royce plc.*)

The mechanical arrangement of the gas turbine, i.e., compressor, combustion, turbine, exhaust, is in series so that the combustion cycle occurs continuously at a constant pressure.

What is a fuel injection system, and what are its advantages and disadvantages?

A fuel injection system delivers metered fuel directly into the induction manifold and then into the combustion chamber (or cylinder of a piston engine) without using a carburetor. Normally, a fuel control unit (FCU) is used to deliver metered fuel to the fuel manifold unit (fuel distributor). From here, a separate fuel line carries fuel to the discharge nozzle in each combustion chamber (or cylinder head in a piston engine, or into the inlet port prior to the inlet valve). With fuel injection, a separate fuel line can provide a correct mixture.

Advantages:

- Freedom from vaporization ice (fuel ice)
- More uniformed delivery of the fuel-air mixture around the combustion chamber section or each cylinder
- Improved control of fuel-air ratio
- Fewer maintenance problems
- Instant acceleration of the engine after idling, i.e., instant response
- Increased engine efficiency

Disadvantages:

- Starting an already hot fuel injection engine may be difficult due to vapor locking in the fuel lines.
- Having very fine fuel lines, fuel injection engines are more susceptible to contamination (i.e., dirt or water) in the fuel.
- Surplus fuel provided by the fuel injection system will pass through a return line, which is usually routed to only one of the fuel tanks. This may result in either the fuel being vented overboard (thus reducing fuel available) or asymmetric (uneven) fuel loading.

What are thrust reverses, and how do they work?

Thrust reverses on jet/gas turbine engine reverse the airflow forward, thereby creating a breaking action. There are two types of thrust reverses: (1) blockers or bucket design and (2) reverse flow through the cascade vane.

Describe maximum takeoff thrust and its limitations.

Maximum takeoff thrust is simply the maximum permissible engine thrust setting for takeoff, expressed either as an N_1 or engine pressure ratio (EPR) figure.

Maximum takeoff thrust is the highest thrust setting of the aircraft's engine when the highest operating loads are placed on the engine. However, as a protection to the engine, maximum takeoff thrust settings have a time limit on their use, namely, 5 minutes for all engines working and 10 minutes with an engine failure.

Note: Some authorities allow a 10-minute time limit with all engines operating.

Describe maximum continuous thrust.

Maximum continuous thrust is simply the maximum permissible engine thrust setting for continuous use, expressed either as an N_1 or engine pressure ratio (EPR) figure.

What is the compression ratio of a gas turbine engine?

The compression ratio of a gas turbine engine is a ratio measure of the change in air pressure between the inlet and outlet parts of either an individual compressor stage or the complete compressor section of the engine.

Individual compressors, either centrifugal or axial-flow types, are placed in series so that the power compression ratio accumulates. For example:

First compressor stage:	Inlet pressure	15 psi	
	Outlet pressure	60 psi	= 4:1
Second compressor stage:	Inlet pressure	60 psi	
	Outlet pressure	180 psi	= 3:1
	Overall compression ratio		12:1

What is the principle of the bypass engine?

The principle of the bypass engine is an extension of the gas turbine engine that permits the use of higher turbine temperatures to increase thrust without a corresponding increase in jet velocity by increasing the air mass/volume intake and discharge to atmosphere via the bypass ducts. Remember,

$$\text{Thrust} = \text{air mass} \times \text{velocity}$$

The bypass engine involves a division or separation of the airflow. Conventionally, all the air entering into the engine is given an initial low compression, and a percentage is then ducted to bypass the engine core. The remainder of the air is delivered to the combustion system in the usual manner. The bypass air is then either mixed with the hot airflow from the engine core in the jet pipe exhaust or immediately after it has been discharged to atmosphere to generate a resulting forward thrust force.

The term *bypass* normally is restricted to engines that mix the hot and cold airflow as a combined exhaust gas. This improves (1) propulsive efficiency and (2) specific fuel consumption and (3) reduces engine noise (this is due to the bypass air lessening the shear effect of the air exhausted through the engine core).

What is bypass ratio?

Bypass ratio in an early single- or twin-spool bypass engine is the ratio of the cool air mass flow passed through the bypass duct to the air mass flow passed through the high-pressure system. Typically, this early evolution of the bypass engine has a low bypass ratio, i.e., 1:1.

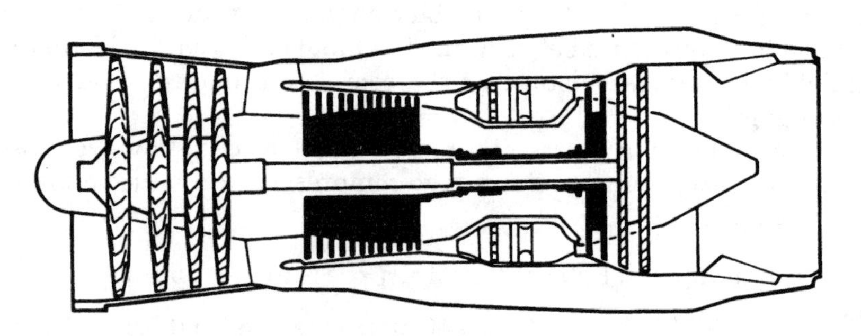

TWIN-SPOOL BYPASS TURBOJET
(low bypass ratio)

Figure 2.7 Bypass airflow for a single- or twin-spool gas turbine engine. (*Reproduced with kind permission of Rolls-Royce plc.*)

Alternatively, bypass ratio for a fan-ducted bypass engines is the ratio of the total airmass flow through the fan stage to the airmass flow that passes through the turbine section/high-pressure (engine core) system. A high bypass ratio, i.e., 5:1, is usually common with ducted fan engines. (*See Figure 2.8, "Triple-spool turbofan engine," page 68.*)

Describe the fan engine and its advantages.

The fan engine can be regarded as an extension of the bypass engine principle (*see Q: What is the bypass ratio? page 66.*) with the difference that it discharges its cold bypass airflow and hot engine core airflow separately.

The turbine-driven fan is in fact a low-pressure axial-flow compressor that provides additional thrust. Normally, the fan is mounted on the front of the engine and is surrounded by ducting that controls the high supersonic airflow speeds experienced at the blade tips, preventing them from suffering from compressibility effect loses.

The fan is either coupled to the front of a number of core compression stages (twin-spool engine), which restricts the width size of the fan, and its bypass air is dusted overboard at the rear of the engine through long ducts, or it is mounted on a separate shaft driven by its own turbine (triple-spool engine), where the bypass airstream is ducted overboard directly behind the fan through short ducts, hence the term *ducted fan*. For example, the CFM56-3 is a twin-spool fan engine, and the Rolls-Royce RB211 is a triple-spool fan engine.

The fan design reflects the specific requirements of the engine's airflow cycle and gives an initial compression to the intake air before it is split between the engine core and the bypass duct. The fan is capable of handling a larger airflow volume than the high-pressure compressor. Therefore, a fan engine normally will have a high bypass ratio, which means that the engine's resulting thrust properties are dominated by the large bypass air mass; therefore, the advantages of the bypass engine are increased further for the fan engine. (*See Q: What are the advantages of a wide-chord fan engine? page 67.*)

The following are some of the main advantages of a fan engine:

1. Smaller engine size.
2. Better propulsive efficiency.
3. Better specific fuel consumption.
4. Reduced engine noise.
5. Contamination (i.e., bird strikes, heavy water) is centrifugally discharged through the bypass duct, therefore protecting the main engine core from damage and even a flameout from water contamination.

What are the advantages of a wide-chord fan engine?

The advantage of a wide-chord fan engine is its ability to optimize the separate thrust properties of the bypass air mass and the engine-core air mass from a single rotating blade row. It does this by using the different speeds between the blade tip and center without using inlet guide vanes, thus keeping engine weight and mechanical complexity to an acceptable level. The chord is the length of the fan's blade from its center mounting to its tip.

It can be said, therefore, that as a basic principle, the wider the fan chord, the higher is the blade tip speed, the greater is the generated airflow speed and mass, and the greater is the discharged air pressure, resulting in an increased engine thrust.

Describe a triple-spool turbofan engine, e.g., the RB211, and its advantages.

A triple-spool turbofan engine such as the Rolls Royce RB211 is a further development of the fan engine that has two distinct differences from the twin-spool fan engine (*see Q: Describe the fan engine and its advantages, page 66*):

1. The triple-spool turbofan engine has three independent compressor spools:
   N_1, the low-pressure compressor spool or fan
   N_2, the intermediate-pressure compressor
   N_3, the high-pressure compressor spool
   and they are each driven by their own turbine and connecting shafts.
2. The front turbofan, or N_1 low-pressure compressor spool, is not connected to any other compression stages.

The turbofan on a triple-spool engine is further improved because it is not restricted to the size of other compressor spools (as it is on a twin-spool engine) and it is driven at its optimal speed by its own turbine. This allows it to have a larger frontal area that consists mainly of a giant ring of large blades, which act more like a shrouded prop than a fan. It is responsible for producing an even larger bypass ratio (i.e., 5:1), which generates approximately 75 percent of the engine's thrust in the form of bypass airflow delivered to the atmosphere via the engine's bypass ducts behind the fan.

The one part of air that flows through the engine N_2 and N_3 compressors becomes highly compressed, of which one-third is used for combustion and two-thirds is used for internal engine cooling.

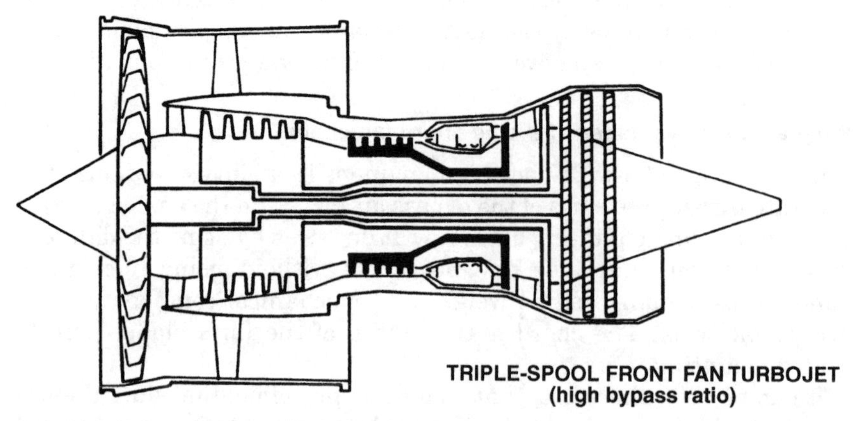

Figure 2.8 Triple-spool turbofan engine. (*Reproduced with kind permission of Rolls-Royce plc.*)

Advantages of the triple-spool fan engine, including the RB211, are twofold:

1. Particular to the triple-spool configuration, including the RB211:
   a. The N_1 fan compressor can be built closer to its optimal design, namely, a wider chord, because it is not restricted by any connection to booster compressors. (*See Q: What are the advantages of a wide- chord fan? page 67.*)
   b. The N_1, N_2, and N_3 compressor sections all work closer to their optimal performance levels because they have their own independent turbine connecting shafts, especially the N_1 spool.
   c. The triple-spool configuration allows more flexibility due to the aerodynamic matching at part load and lower inertia of the rotating parts.
   d. Higher engine thrust output due to the improved fan section (point *a*) and the improved independent spool configuration (points *b* and *c*).
   e. Easier to start because only one spool needs to be turned by the starter.
   f. The triple spool's modular assembly makes it easier to build and in particular to maintain; i.e., if the N_3 compressor suffers a fault, then the modular assembly of the engine allows for the N_3 section of the engine alone to be removed for repair.
2. Advantages common to fan engines. (*See Q: Describe the fan engine and its advantages, page 66.*)

Why is a fan engine flat rated?

The fan engine is flat rated to give it the widest possible range of operation, keeping within its defined structural limits, especially in dense air.

Note: Flat rating guarantees a constant rate of thrust at a fixed temperature.

When and where is a jet/gas turbine (bypass) engine at its most efficient, and why?

At high altitudes and high rpm speeds. (*See Q: Why does a jet aircraft climb as high as possible? page 69.*)

Why does a jet aircraft climb as high as possible?

Jet aircraft climb as high as possible (i.e., to their service ceiling) because the gas turbine (bypass) engines are most efficient when their compressors are operating at high rpms, approximately 90 to 95 percent. This high rpm speed results in the engine's optimal gas flow condition that achieves its best specific fuel consumption (SFC). However, this high rpm speed can be achieved only at high altitudes because only at high altitudes where the air density is low will the thrust produced be low enough to equal the required cruising thrust.

The primary reason for designing an engine's optimal operating condition at approximately 90 to 95 percent rpms is to make it coincident with the best operating conditions of the airframe, namely, minimum cruise drag.

Therefore, at high altitudes there are two main consequences:

1. *Minimum cruise airframe drag.* This is experienced at high altitudes because drag varies only with equivalent airspeed (EAS), i.e., as EAS decreases, drag decreases. At very high altitudes, i.e., above 26,000 ft, the Mach number (MN) speed becomes limiting, and therefore, EAS and true airspeed (TAS) are reduced for a constant MN with an increase in altitude. (*See Q: Describe Mach number, page 122.*) Therefore, the lowest cruise EAS is at the highest attainable altitude (service ceiling), and because drag varies only with EAS, airframe drag is also at its lowest value at high altitudes. Consequently, our thrust requirements are lower at high altitudes because our thrust value must only be equal to our drag value.
2. *Best engine SFC.* This is experienced at high altitudes due to the engine's ability to operate at its optimal high-rpm condition because of the low atmospheric air density. The engine's best SFC operating conditions are a function of its internal aerodynamic design. This reflects the optimization of the engine to be generally at its best under conditions where it will spend most of its working life, i.e., high altitudes, high-speed conditions, and comparatively high engine rpm settings.

High-altitude conditions optimize the engine's design by using the reduction in the atmospheric air density as a reduced airflow mass into the engine for a given engine rpm speed with an increase in altitude. The fuel control system adjusts the fuel delivery to match the reduced mass airflow to maintain a constant mixture and so maintain a constant engine speed. This causes the thrust to fall for a given rpm speed and requires an increase in compressor rpms to maintain its thrust values with an increase in altitude until its optimal high-rpm speed is reached.

In addition, the required thrust is lower with an increase in altitude because the EAS and airframe cruise drag reduce with altitude. Therefore, it follows that only at high altitudes will the thrust be low enough to equal the required thrust at the engine's optimal normal cruising high rpm that achieves its best SFC. (*See Q: What advantages does a jet-engined aircraft gain from flying at a high altitude? page 71.*)

What advantages does a jet-engined aircraft gain from flying at a high altitude?

The advantages a jet engine gains from flying at high altitudes are

1. Best specific fuel consumption (SFC)/increased (maximum) endurance

Note: Endurance is the need to stay airborne for as long a time as possible for a given quantity of fuel. Therefore, the lowest SFC in terms of pounds of fuel per hour is required.

2. Higher true airspeed (TAS) for a constant indicated airspeed (IAS), providing an increased (maximum) attainable range

Note: Maximum attainable range is the greatest distance over the ground flown for a given quantity of fuel, or the maximum air miles per gallon of fuel.

1. Best SFC and thus increased (maximum) endurance are achieved at high altitudes because of two effects:
   a. Minimum cruise drag is experienced at high altitudes because the Mach number (MN) speed becomes limiting above approximately 26,000 ft, and for a constant MN (as is the normal operating practice), the TAS and equivalent airspeed (EAS) decrease with altitude, and drag varies only with EAS. As such, the EAS is reduced progressively to a level closer to the aircraft's best endurance speed, the higher the altitude, which is obviously where drag is least.

Note: Minimum drag speed (VIMD), broadly speaking, remains constant with altitude. And because minimum aircraft drag requires minimum thrust (i.e., thrust = drag), and given that thrust is a product of engine power and fuel consumption is a function of engine power used, the aircraft thus has its lowest fuel consumption in terms of fuel used per hour. Hence it produces the maximum endurance flight time for a given quantity of fuel when flying at its lowest cruise EAS (best endurance speed), which the highest attainable altitude will achieve for a constant MN.

   b. In addition, an aircraft's fuel consumption decreases slightly at high altitudes because of the higher propulsive efficiency of the engine (*see preceding question*), which brings it closer to its best SFC operating condition. Therefore, an aircraft's best SFC and maximum endurance are attained by flying at
   (1) The best endurance speed (VIMD)
   (2) Highest possible altitude where the engine achieves its best propulsive efficiency
2. The higher the altitude, the greater is the TAS for a constant IAS, which provides an increased (maximum) attainable range.

Maximum range is also defined by an EAS that is a slightly higher speed than the best endurance speed because the benefits of the increased IAS (i.e., greater ground speed/distance covered) outweighs the associated increased drag and higher fuel consumption. (*See Q: Define maximum endurance and range with reference to the drag curve, page 209.*)

The higher TAS for a constant IAS results simply from the reduction in air density at higher altitudes. Thus ground speed/ground distance covered and range are increased (or the flight time is reduced for a given distance) at high altitudes for a constant IAS that gives a higher TAS.

Thus the rule of thumb for best range is: The higher the better. Just how high depends on other factors, such as winds at different levels and sector lengths.

Note: Normal operating practice for jet aircraft will be to have a limiting MN speed above approximately 26,000 ft. Therefore, a constant MN is flown above 26,000 ft, which would see TAS decrease with altitude because IAS and the local speed of sound (LSS) decrease to maintain a constant MN. (*See Qs: Describe local speed of sound (LSS) and Mach number (MN), page 122; How does temperature affect local speed of sound (LSS), page 122; A flight carried out below optimal altitude has what result on jet performance? page 211.*)

Therefore, because TAS decreases at high altitude for a constant MN, this results in the ground speed being reduced slightly, and thus obtainable range also will be reduced slightly below its maximum range and/or its flight time will increase for still-air conditions. To counter this, a slightly higher long-range cruise MN speed (and thus IAS) can be selected that increases the TAS for the altitude, although this would be detrimental to endurance.

In basic terms of best SFC and endurance and greater TAS and range, an aircraft should remain as high as possible for as long as possible. (*See Q: A flight carried out its below optimal altitude has what result on a jet's performance? page 211.*)

Explain the jet/gas turbine engine's thrust-to-thrust lever position.

The thrust lever produces more engine thrust from its movement near the top of its range than the bottom.

An engine's operating cycle and gas flow are designed to be at their most efficient at a high-rpm speed, where it is designed to spend most of its life. (*See Q: When and where is a jet/gas turbine engine at its most efficient and why? page 69.*) Therefore, as rpms rise, mass flow, temperature, and compressor efficiency all increase, and as a result, more thrust is produced, say, per 100 rpm, near the top of the thrust lever range than near the bottom.

In practical terms, this translates to differing thrust output per inch of thrust lever movement; i.e., at low-rpm speed (near the bottom of its range), an inch movement of the thrust lever could produce only 600 lb of thrust, but at a high-rpm speed (near the top of its range), an inch movement of the thrust lever typically could produce 6000 lb of thrust.

For this reason, if more power is required at a low thrust lever setting, then a relatively large movement/opening of the thrust levers is required, i.e., when initiating a go-around/overshoot. However, if operating at the top of the thrust lever setting, a large reduction or increase in thrust would only require a relatively small movement of the thrust lever.

Obviously, an appreciation of the jet/gas turbine engine's response characteristics helps the pilot's understanding and operation of his or her engines.

What are the main engine instruments?

The main primary engine instruments usually are

1. Engine pressure ratio (EPR) gauge (thrust measurement)
2. N_1 gauge (low compressor rpms)
3. Exhaust gas temperature (EGT) or total gas temperature (TGT) (engine temperature)

Other possible primary engine instruments are

4. N_2 gauge (intermediate compressor rpms)
5. Fuel flow (fuel flow indicator)

Secondary engine instruments usually are

6. Oil temperature gauge, pressure gauge, and quantity gauge
7. Engine vibration meter

What is engine pressure ratio (EPR)?

Engine pressure ratio (EPR) is the ratio of air pressure measurements taken from two or three different engine probes and displayed on the EPR gauge for the pilot to use as a parameter for setting engine thrust.

The EPR reading is the primary engine thrust instrument, with the temperature of the turbine stage governing the engine's maximum attainable thrust.

Normally, EPR on a gas turbine–powered aircraft is a ratio measurement of the jet pipe pressure to compressor inlet ambient pressure or sometimes the maximum compressor cycle pressure to compressor inlet ambient pressure. However, on a fan engine, the EPR is normally a more complex ratio measurement of an integrated turbine discharge and fan outlet pressure to compressor inlet pressure.

What is exhaust gas temperature (EGT), and why is it an important engine parameter?

EGT is exhaust gas temperature and is an important engine parameter because it is a measure/indication of the temperatures being experienced by the turbine.

The only real operating threat to the engine's life is excessive turbine temperatures. The maximum temperature at the turbine is critical because if the EGT limit is exceeded grossly on startup (hot start), the excessive temperatures will damage the engine, especially the turbine blades. Also, if the cruise EGT limit is exceeded slightly for a prolonged period, this will shorten the engine's life.

Describe an engine wet start and its causes, indications, and actions.

An engine wet start is otherwise known as a failure to start after the fuel has been delivered to the engine. The cause of a wet start normally is an ignition problem.

Indications of a wet start are

1. Exhaust gas temperature (EGT) does not rise.
2. Revolutions per minute (rpms) stabilize at starter maximum.

Actions required for a wet start include

1. Close the fuel lever/supply as soon as a wet start is diagnosed (usually at the end of the starter cycle).
2. Motor over the engine to blow out the fuel (approximately 60 seconds).

Describe an engine hung start and its causes, indications, and actions.

A hung start occurs when the engine ignites but does not reach its self-sustaining rpms. (Self-sustaining speed is an rpm engine speed at and above which the engine can accelerate on its own without the aid of the starter motor.)

The cause of a hung start is insufficient airflow to support combustion due to the compressor not supplying enough air because of one or a combination of the following:

1. High altitude, low-density air
2. Hot conditions, low-density air
3. Inefficient compression
4. Low starter rpms

Indications of a hung start include

1. High exhaust gas temperature (EGT), above normal
2. Engine rpms below normal self-sustaining speed

Actions required for a hung start are

1. Close fuel lever/stop fuel delivery.
2. Motor over the engine to blow out the fuel (for approximately 60 seconds)

Note: To gain a successful start in hot and high conditions, you have to introduce more air into the engine. Adjusting the fuel supply does not help.

Increasing fuel = rpm decrease and EGT increase

Decreasing fuel = rpm increase and EGT decrease

Describe an engine hot start and its causes, indications, and actions.

A hot start is one in which the engine ignites and reaches self-sustaining rpms, but the combustion is unstable and the exhaust gas temperature (EGT) rises rapidly past its maximum limit.

Causes of a hot start include

1. Overfueling (throttle open)
2. Air intake/exhaust blocked
3. Tailwind, causing the compressor to run backward
4. Seized engine, e.g., ice blockage

Indication of a hot start is an EGT rising rapidly toward its maximum limit.

Actions required for a hot start are

1. Close fuel lever/stop fuel delivery before the EGT limit has been reached.
2. When the engine rpms have slowed to the reengagement speed, motor over the engine to blow out the fuel (approximately 60 seconds).

What is a variable/reduced-thrust takeoff?

A variable/reduced-thrust takeoff uses the takeoff thrust (EPR/N_1) required for the *aircraft's* actual takeoff weight, which is a reduced thrust value from the maximum takeoff weight thrust value that meets the aircraft's takeoff and climb performance requirements with one engine inoperative.

The full takeoff thrust is calculated against an aircraft's performance-limited (either field length, WAT, tire or net flight path, obstacle-clearance climb profile) maximum permissible takeoff weight and not its actual takeoff weight. The reduced takeoff thrust is the correct thrust setting for the aircraft's actual takeoff weight that achieves the aircraft's takeoff and climb, one engine inoperative, performance requirements.

A reduced/variable takeoff thrust is calculated by using the assumed/flexible temperature performance technique. (*See Q: What is an assumed/flexible temperature? page 201.*)

Using this variable takeoff engine pressure ratio (EPR) means that the aircraft is now operating at or near a performance-limiting condition, whereby following an engine failure the whole takeoff would still be good enough in terms of performance. In fact, a lower-weight aircraft using a variable thrust will mirror the takeoff run/profile of a maximum takeoff weight aircraft using a full thrust setting.

Note: A function of a reduced thrust (assumed/flexible) takeoff is as follows: A higher assumed/flexible temperature relates to a lower air density delivered to the engine. The fuel flow is reduced correspondingly to maintain the correct air-fuel mixture. This produces a lower theoretical thrust from the engines, and practically, this is even lower because the actual air density is greater and therefore the mix is air dominant and less explosive. This lower thrust therefore makes the aircraft's momentum more dominant in reaching its V1 and VR speeds. This greater dependence on momentum requires a longer runway (TOR/D) for a given weight, or it relates to the same takeoff run (TOR) performance (i.e., rotation point) for a lower-weight aircraft using a derated takeoff thrust against a maximum takeoff weight aircraft using a maximum thrust setting.

A variable/reduced-thrust takeoff can be employed on most occasions when an aircraft's takeoff weight is lower than the performance-limited maximum takeoff weight for the ambient air temperature. Its use is optional, but the higher EPR required for the maximum takeoff weight always must be available whenever a variable/reduced-thrust takeoff is being conducted. It is recommended that full takeoff power be restored in the event of an engine failure above V_1 or whenever deemed necessary. The use of maximum takeoff weight (MTOW) EPR for an aircraft with a lower actual weight obviously will provide the aircraft with a better all-round takeoff performance.

It should be clearly understood that a reduced/variable takeoff is a reduced-thrust takeoff in relation to the thrust setting required for the maximum permissible takeoff weight. It is in fact the correct thrust takeoff setting for the aircraft's actual weight that meets its performance requirements.

Can a maximum takeoff weight aircraft use a reduced takeoff technique?

Yes, a reduced-thrust takeoff can be used even when an aircraft is at its maximum takeoff structural weight, providing the TOR/D is not limiting. This is so because you can trade momentum gained from a longer TOR/D to achieve the V_1 and V_R speeds at the performance-limiting conditions for a lower thrust setting.

Why do you use reduced derated thrust takeoffs in a jet aircraft?

There are two main reasons for using a reduced derated takeoff.

1. *To protect engine life and to improve engine reliability.* Derated thrust takeoffs reduce the stress and attrition of the engine during the takeoff period when the highest loads are placed on the engine.
2. *To reduce the noise generated by the aircraft.* (Noise suppression of this type normally is used for takeoff and occasionally on approaches over noise abatement areas, as well as for nighttime flying noise restriction.

Why is the risk per flight decreased with a reduced-thrust takeoff?

When a reduced thrust is used for takeoff, the risk per flight is decreased because of the following main reasons:

1. The assumed/flexible temperature method of reducing thrust to match the takeoff weight does so at a constant thrust-weight ratio, making the actual takeoff distance and takeoff run distance from the reduced-thrust setting less than that at full thrust and full weight by approximately 1 percent for every 3°C that the actual temperature is below the assumed temperature.
2. The acceleration-stop distance is further improved by the increased effectiveness of full-reverse thrust at the lower temperature.
3. The continued takeoff after engine failure is protected by the ability to restore full power on the operative engine.

What are the limitations of a variable/reduced-thrust (flex) takeoff?

Clearly, a variable/reduced-thrust (flex) takeoff can only be used when full takeoff thrust is not required to meet the various performance requirements on the takeoff and initial climb out. Therefore, the limitations of using a reduced-thrust takeoff are

1. Not maximum takeoff weight limited by
   a. Takeoff field length
   b. Takeoff weight-altitude-temperture (WAT) curve (engine-out climb gradient at takeoff thrust)

c. Net takeoff flight path (engine-out obstacle clearance)
2. Maximum outside air temperature (OAT) limitation

Where the proposed takeoff weight is such that none of the preceding considerations are limiting, then the takeoff thrust may be reduced until one of the considerations listed below becomes limiting.

Note: Based on one engine inoperative.

3. Furthermore, derated (flex) thrust takeoffs are limited by
   a. The derated (flex) thrust must not be reduced by more than a set amount (engine-specific), e.g., 25 percent below the maximum takeoff weight full-rated takeoff thrust.

Note: This is normally restricted by a maximum flex or assumed temperature constraint, i.e., T_{max}, flex. That is,

$$\text{Flex temp} < T_{max}$$

   b. The derated (flex) temperature cannot be lower than T_{ref} (flat rating cutoff temperature that guarantees a constant rate of thrust at a fixed temperature, which equates to the climb thrust); i.e., flex temp > T_{ref}.
   c. The derated (flex) thrust cannot be lower than the maximum continuous thrust used for the final takeoff flight path.
   d. The actual outside air temperature (OAT); i.e., flex temp > OAT. (*See Q: What is an assumed/flexible temperature? page 201.*)
4. Derated/flexible takeoffs should not be used on the following occasions
   a. On an icy or very slippery runway.
   b. On contaminated runways, i.e., precipitation covered, etc.
   c. When reverse thrust is inoperative.
   d. When the antiskid system is inoperative.
   e. When an increased V_2 procedure is used in order to improve an obstacle-limited takeoff weight.

What happens to engine pressure ratio (EPR) on the takeoff roll?

For the purpose of this question, let us assume that EPR is a measure of the jet pipe exhaust pressure (P_7) against compressor inlet ambient pressure (P_2). Prior to opening the throttle levers, the EPR reading will be very low, if not a 1:1 ratio. Now, when the throttle levers are advanced, the EPR reading will decrease initially because the P_2 pressure increases against a constant or slowly increasing P_7 value; therefore, the ratio decreases before steadily increasing to its takeoff setting.

There are two reasons for this. First, the engine compressor, combustion, and turbine stages are in series, and this leads the engine to suffer from a slow response (or lag) to a throttle input. This is so because the greater amount of air induced into the engine takes time to move through the compressor, combustion, and turbine stages before it is expelled from the engine with a resultant/reaction forward force.

Second, a consequence of the engine's slow response rate or lag is its effect on the EPR reading because the reading of the (P_7) jet pipe exhaust pressure is taken from the rear of the engine and the (P_2) compressor inlet pressure reading is taken from the front of the engine. Therefore, as the engine throttles are opened up, initially the compressor inlet air pressure (psi) will increase before the jet pipe exhaust air pressure (psi) increases proportionally, thus creating an increased or initially quickly increasing P_2 value for a constant or slowly increasing P_7 value. This results in an initial decrease in the EPR reading. Then, as the engine accelerates along its entire length, the engine turbine and compressor rpm speeds increase, and the EPR reading increases steadily to its takeoff setting. For example:

Idle	P_7: 40 psi/P_2: 38 psi	= 1.05 EPR
Initial lag	P_7: 40 psi/P_2: 45 psi	= 0.89 EPR
Steady increase	P_7: 60 psi/P_2: 48 psi	= 1.25 EPR
Takeoff power	P_7: 120 psi/P_2: 60 psi	= 2.00 EPR

Why does engine pressure ratio (EPR) need to be set by 40 to 80 knots on the takeoff role?

The EPR for takeoff has to be set by 40 to 80 knots (the exact speed is type-specific) for the following main reasons:

1. So that the pilot is not chasing rpm needles on the takeoff roll.
2. To ensure an adequate aircraft acceleration so that the performance-calculated V_1 and V_R speeds are achieved by the takeoff run required (TORR) rotate point for the given aircraft weight and ambient conditions.

What is an engine windmill start, and when is it used?

A windmill start occurs when the engine is started without the aid of the starter because the compressors are being turned by a natural airflow when airborne. This delivers the air charge to the combustion chambers, where fuel and an ignition spark are introduced as normal for a stable engine relight. This effect is known as *windmilling,* and as such, windmill starts are used to relight an engine when airborne. (*See Q: What is the purpose of engine relight boundaries? page 79.*)

What is the purpose of engine relight boundaries?

The purpose of engine relight boundaries is to ensure that the correct proportion of air is delivered to the engine's combustion chamber to restart the engine in flight. For this reason, the aircraft's flight manual outlines the approved relight envelope of airspeed against height. This ensures that within the limits of the envelope, the airflow ingested into the engine will rotate the compressor at a speed that generates and delivers a sufficient volume of air into the combustion chamber to relight the engine successfully.

Note: The approved flight envelope usually will be subdivided into starter assist and windmill boundaries.

What causes a jet/gas turbine upset, and how do you correct it?

Disturbed or turbulent airflow will cause a jet/gas turbine engine to be upset and to stall.

This occurs because a jet/gas turbine engine is designed to operate using a clean uniform airflow pattern that it obtains within the aircraft's normal operating attitude. However, beyond the aircraft's normal angles of incidence and slip and/or in extremely severe weather turbulence, the engines can experience a variation in the ingested air's pressure/density, volume, angle of attack, and velocity properties. This changes the incidence of the air onto the compressor blades, causing the airflow over the blades to break down and/or inducing aerodynamic vibration. This upsets the operation of the engine causing it to stall.

The stall can be identified by (1) increases in total gas temperature (TGT), (2) engine vibration, and (3) rpm fluctuations.

What is a jet engine surge, what causes it, and what are the indications?

A *surge* is the reversal of airflow through an engine, where the high-pressure air in the combustion chamber is expelled forward through the compressors, with a loud bang and a resulting loss of engine thrust.

A surge is caused when

1. All the compressor stages have stalled, e.g., bunt negative-*g* maneuver.
2. An excessive fuel flow creates a high pressure in the rear of the engine. The engine will then demand a pressure rise from the compressors to maintain its equilibrium, but when the pressure rise demanded is greater than the compressor blades can sustain, a surge occurs, creating an instantaneous breakdown of the flow through the machine.

A surge is indicated by

1. Total loss of thrust.
2. A large increase in TGT.

The required actions in response to an engine surge are

1. Close the throttles smoothly and slowly.
2. Adjust the aircraft's attitude to unstall the engines, which lead to the surge.
3. Slowly and smoothly reopen the throttles.

Why are bleed valves fitted to gas turbine engines?

Bleed valves are fitted on gas turbine engines for two main reasons

1. To provide bleed (tap) air for auxiliary systems. For example:
   a. Air conditioning and cabin heating/pressurization/EFIS cooling/cargo heating
   b. Engine cooling, especially
      (1) The combustion chamber
      (2) The turbine section
   c. Accessory cooling (generator, gearbox, and other engine-driven systems)
   d. Engine and wing anti-icing systems
1. To regulate the correct airflow pressures between different engine sections.

Why do gas turbine engines have auto igniters, and how do they work?

Auto igniters are used in gas turbine engines to protect against disturbed/turbulent airflow upsetting the engine. This condition is particularly common with rear-mounted engines during some abnormal and even some rather normal flight maneuvers because rear-mounted engines are placed ideally to catch any disturbed airflow generated by the wing when the airflow pattern brakes down as a result of either a high incidence of attack (e.g., prestall buffet), high-*g* maneuvers (e.g., steep turns), or high Mach number effect.

Auto igniters work by sensing a particular value of incidence of the aircraft, via the incidence-sensing (probe) system (which is also used to activate the stick shaker and pusher), and automatically signals on the ignition system before the disturbed airflow generated by the wing affects the engines, thus ensuring that the engines at least continue to run, although in some cases they might surge a little.

What is FADEC?

FADEC is full authority digital engine control and is a system that automatically controls engine functions, i.e., start procedures, engine monitoring, fuel flow, ignition system, and power levels required.

FADEC computers can be found on the A320 aircraft's engine and on the EJ200 engines used on military aircraft.

What fuels are used commonly for civil jet aircraft?

The fuels used for gas turbine civil aircraft engines are

1. *Jet A1 (Avtar).* This is a kerosene-type of fuel with a normal specific gravity (SG) of 0.8 at 15°C. It has a medium flash point and calorific value, a boiling range of between 150 and 300°C, and a waxing point of −50°C.
2. *Jet A.* This is similar to A1, but its freezing point is only −40°C.

Note: Jet A normally is available only in the United States.

(*For fuel questions, see Chapter 6, "Aircraft Instruments and System," pages 172–174.*)

How do jet/gas turbine engines generate noise?

The noise generated by a jet/gas turbine engine is from the sheer effect of different displaced air velocities. The sheer is the difference between the jet's faster displaced air and the slower ambient air around it.

How is jet/gas turbine engine noise controlled or reduced?

Jet/gas turbine engine noise can be reduced by the following:

1. Bypass engines. This reduces the sheer effect between the displaced slower bypass engine air and the ambient air that reduces noise.
2. Reduced-thrust takeoff.

Jet/gas turbine engine noise can be controlled by the following:

3. Maximum-angle climb after takeoff. This allows the aircraft to get above any noise-control zones.

Is there a critical engine on a jet/gas turbine aircraft?

There is no critical gas turbine engine because the engines are positioned symmetrically with opposing revolution direction. However, there is a governor engine, i.e., an engine that is the master that sets the rpm speed for the others. (*See Q: How does a crosswind affect the critical engine? page 60.*)

Engine Fire Detection and Protection Systems

Describe a typical aircraft fire detection and protection system.

A typical engine fire detection and protection system would consist of the following:

1. Overheat and fire detection loop(s) with multisensors. Note that normally a minimum of two separate systems (loops) exist per unit. These are coupled with visual (lights) and aural indications for both overheat and fire detected conditions for
   a. Each engine
   b. Auxiliary power units (APUs)
   c. Wheel wells (not on all aircraft types)
2. Fault monitoring system of overheat and fire detection systems. This is coupled with visual indications for
   a. Engines
   b. APUs
3. Fire extinguishers and firing circuits for
   a. All engines. Note that usually there are a minimum of two extinguisher bottles that can fire into each engine so as to provide a second extinguishing supply to all engines. This is coupled with visual indications of bottle discharge.
   b. APUs. Note that usually there is a separate single fire extinguisher bottle. This is coupled with visual indications of bottle discharge.
4. Testing facility of
   a. Firing circuits for
      (1) Engine fire extinguisher(s)
      (2) APU fire extinguisher
   b. Fault monitoring system
   c. Overheat/fire detection loops for
      (1) Each engine
      (2) APUs
5. Lavatory/cargo holds smoke detection system.

What are the indications of thermal expansion and use of the fire bottles on the side of the aircraft fuselage?

Separate disks, i.e., one for extinguishent release due to thermal expansion and one to indicate use, normally are found on the side of the aircraft's fuselage. If they are intact, they indicate that the extinguishent is still in the fire bottle.

Note: The color and location of such disks are aircraft type-specific.

Describe a typical engine fire drill.

(*See Chapter 9, "Flight Operations and Technique," page 288.*)

Chapter

3

Jet and Propeller Aircraft Differences

Describe the approach differences between a jet and a piston engine propeller aircraft.

There are six main handling differences between a jet aircraft and a propeller aircraft on the landing approach. In all the differences, the jet aircraft is worse off than the propeller aircraft in maintaining the approach profile and in correcting errors on the approach.

The six main differences involve aerodynamics and the engines. The aerodynamic differences include

1. *Momentum.* The jet aircraft has a greater momentum than the lighter and slower propeller-driven aircraft. Therefore, the responses of a jet aircraft to changes in flight path are much slower, and sudden changes are virtually impossible.
2. *Speed stability.* The jet aircraft suffers from poor speed stability being an inducement to a low-speed condition.
3. *Wing lift values.* The jet aircraft's swept wing produces less lift than the propeller aircraft's straight wing. The swept wing experiences a faster increasing drag penalty than lift, resulting in a high sink rate at low speeds.

The engine differences include

4. *Engine response rate/acceleration and deceleration.* The jet engine has a poorer acceleration response at low-rpm speeds known as *lag*. The propeller drag produces a decelerating braking

action (which can be very useful on the approach) but is absent on the jet aircraft.

5. *Slipstream effects.* The propeller slipstream produces an immediate extra lift value over the wings at a constant airspeed but is absent on jet aircraft.

6. *Power-on stall speed.* The stall speed is significantly lower when the engine power is increased or on for a propeller aircraft because it generates a slipstream with an increased airflow speed over the wing that increases the lift produced and thus reduces the aircraft's effective weight, also reducing the stall speed. For a jet aircraft, which does not produce a slipstream over the wing, the stall speed is virtually unchanged with its power on.

Explain the (low) speed control difference between a jet aircraft and a propeller-driven aircraft.

The speed control for both jet and propeller-driven aircraft is a function of drag against speed. The drag experienced by both aircraft has the same dynamic qualities in that to balance drag against a speed, the thrust/power must be set to a corresponding required value. However, the value of drag against a speed is different between jet and propeller-driven aircraft, and it is this difference that makes their speed control properties markedly different. (*See Q: Describe the drag curve on a piston-engined/jet aircraft, pages 7 and 8.*)

The jet aircraft's relatively flat drag curve over the low speed range makes it difficult to select the correct thrust for the required speed because of the minimal difference in thrust required values across a large speed range. Therefore, it is easy to select an incorrect speed control thrust setting, and any slight difference, whether higher or lower, will set up a speed divergence that is also difficult to detect. A speed divergence due to poor speed control on jet aircraft can result in the aircraft's speed slipping into an unstable region where the aircraft suffers from poor speed stability. (*See Q: What speed stability differences are there between a jet aircraft and a propeller-driven aircraft? page 87.*)

In contrast, the propeller-driven aircraft's drag curve is markedly different across its low-speed range, and therefore, it is much easier to determine and set the correct power to balance the drag and to deliver the desired speed. Also, incorrect settings give rise to noticeable handling characteristics and more marked speed divergences that are very noticeable to the pilot, in stark contrast to the jet aircraft's qualities. Therefore, the speed control of the propeller-driven aircraft is easier to manage.

What speed stability differences are there between a jet aircraft and a propeller-driven aircraft?

The jet aircraft's speed stability is much poorer than that of a propeller-driven aircraft because of the following two main differences:

1. The jet's recommended threshold speed (i.e., $1.4V_S$) tends to be in the neutral or unstable speed range.
2. Thrust changes with speed, which helps to improve the speed stability on propeller-driven aircraft and does not improve it on jet aircraft. (*See Q: Explain speed stability, page 22.*)

Describe the difference between a jet and a piston-engined propeller aircraft's stall speed.

The stall speed of a piston-engined propeller aircraft generally is a slower speed than that of a jet aircraft. Also, the range of the stall speeds is much larger for a jet aircraft than for a piston-engined propeller aircraft.

Describe the differences between propeller and jet aircraft wing performance.

The performance margin of a propeller-driven aircraft's straight wing is greater than that of a jet aircraft's swept wing, especially when contaminated. The reason for the improved relative performance on a propeller-driven aircraft is that the wing is unswept and there is a high-energy airflow (prop wash) over the upper surface of the wing that partly offsets some of the contaminant effects.

On a jet, the available performance margin with a contaminated wing is practically zero for two main reasons. First, there is no upper-surface high-energy flow available. Second, the wing sweep decreases the amount of lift generated for a given angle of attack relative to the straight wing so that higher angles of attack are required from the swept wings, which in turn results in greater performance sensitivity to wing contamination.

Chapter

4

Navigation

Chart Navigation

What is true direction?

True direction is measured with reference to true north.

What is magnetic direction?

Magnetic direction is measured with reference to magnetic north.

Note: The magnetic north pole is near to the geographic true north pole but is not in exactly the same place.

What is magnetic variation?

Variation is the difference between the direction of magnetic north and true north. It is therefore the difference between all true headings and their corresponding magnetic headings at any point. Variation is not a constant across the earth but varies in magnitude from place to place. Variation is described as either east (plus) or west (minus).

What is compass direction?

Compass direction is measured with reference to compass north. Compass north is the deviated direction away from magnetic north indicated by a compass needle. (*See Q: What is compass deviation? page 89.*)

What is compass deviation?

Deviation is the difference between the direction of magnetic north and compass north.

What are the lines that run from pole to pole on the earth called?

Meridians of longitude run from pole to pole and converge at both poles.

Meridians of longitude are described by their angle away from the prime meridian, namely, the Greenwich meridian. The units of measurement used to describe a meridian's position are degrees, minutes, and seconds, and they are named east or west depending on where they lie in relation to the prime meridian. The maximum longitude of 180 degrees is known as the Greenwich antimeridian. Additionally, meridians at 15-degree intervals in an easterly direction are classified as alpha to zulu. The Greenwich meridian is known as zulu.

How do you measure a change of latitude?

A change of latitude, i.e., north-south distances, along a meridian of longitude is measured as 1 nautical mile equals 1 minute (60 minutes = 1 degree).

Note: North-south distances can be measured across a pole where points lie on opposite meridians of longitude.

What are the lines that run east-west on the earth called?

Parallels of latitude run east-west on the earth.

Parallels of latitude are described by their angle above or below the equator. The units of measurement used to describe a parallel of latitudes position are degrees, minutes, and seconds (1 degree = 60 minutes, 1 minute = 60 seconds). The maximum possible latitude is 90 degrees north or south, i.e., north or south pole.

How do you measure a change of longitude?

A change of longitude, i.e., east-west distance, (along a parallel of latitude) has a varied distance (nautical miles) per degree at different latitudes. The distance along the equator is measured as 1 nautical mile equal to 1 minute. As we move away from the equator, the east-west distance (nautical miles) decreases for the same change in longitude until eventually, at the poles, the distance between the meridians has reduced to zero. Therefore, the change of longitude distance decreases as latitude increases, and this is known as *departure*. (*See Q: What is departure? page 90.*)

What is departure?

Departure is an east-west distance along a parallel line of latitude, other than the equator, that requires the use of the following formula to calculate the variable east-west distance (nautical miles) for a given change of longitude at different latitudes.

Departure (nautical miles) = change of longitude (minutes) × cosine of latitude

What is a great circle track?

A great circle track is a line of shortest distance between two points on a sphere (or a flat surface) with a constantly changing track direction as a result of convergence.

What is a rhumb line?

Rhumb lines are tracks with a constant track direction between two points on a sphere and therefore must be a longer distance than a great circle track.

What is convergency?

Convergency represents the change of direction experienced along east-west tracks, except rhumb lines, as a result of the way direction is measured due to the effects of converging meridians at the poles. The change of direction experienced between two points is known as *convergency.*

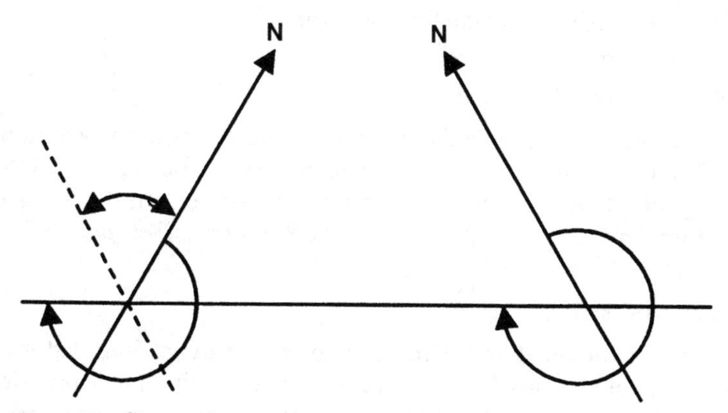

Figure 4.1 Convergency.

Convergency is clearly dependent on latitude; it is zero at the equator, where the meridians are parallel, and a maximum at the poles, where the meridians converge. It is also dependent on how far you travel. A short distance traveled will have only a small change of direction, whereas a long distance will have a large change of direction.

What is a conversion angle?

Conversion angle is the angle of difference in direction between the rhumb line and the great circle track between two points and therefore is used to calculate a rhumb line direction given a great circle track, or vice versa.

What is chart scale?

Scale is defined as the ratio of chart distance to earth distance.

Airborne Navigation Equipment

Describe the doppler effect.

Doppler effect is the change of frequency between the transmitted and received signals, known as *Doppler shift,* due to movement of the transmitter. This effect is present in radio waves and, in particular, radar.

With a static transmitter and a static receiver, the received frequency is the same as the transmitted frequency. If the transmitter is moving toward the receiver, more cycles are received every second; i.e., the frequency increases as one moves toward the receiver. If the transmitter is moving away from the receiver, fewer cycles are received every second; i.e., the frequency decreases as one moves away from the receiver. The classic example of the Doppler effect is the change in pitch of a train's whistle as it passes a stationary observer.

What is an airborne Doppler system?

Doppler is a self-contained onboard radio/radar navigation system based on the Doppler effect principle that operates in the 8.8- to 13.2-GHz frequency bands and is used to mathematically calculate the ground speed of an aircraft. (*See Q: Describe the Doppler effect. page 92.*)

What is an INS/IRS?

An INS is an onboard self-contained inertia navigation system that can provide continuous information on an aircraft's position without any external assistance. An IRS (inertia reference system) is a modern INS that usually has a greater integration into the flight management system (FMS), and provides the aircraft's actual magnetic position and heading information with reference to the FMS required position and heading.

The directional acceleration information provided from the INS's accelerometers and gyroscopes is calculated by the position computer that determines the aircraft's present latitude and longitude position, provided a correct initial position has been input.

How does an INS/IRS work?

The general principle of all inertia navigation systems (INSs) is that the system measures the aircraft's inertia movement from an initial position as a great circle track direction and distance to continuously determine its up-to-date position.

The components of an INS are

1. Accelerometers
2. Gyroscopes
3. Position computer

The aircraft moves in three dimensions, but the navigation equipment is only interested in acceleration in the horizontal plane. Therefore, the key to the whole INS arrangement is the *accelerometers*.

How does an INS/IRS find true north?

An inertia navigation system (INS) is aligned to true north by its gyroscopes.

How does an INS/IRS find magnetic north?

INS/IRS systems find magnetic north by applying a stored magnetic variation to the calculated true north (*see preceding question*).

What are the advantages of an INS?

The advantages of an inertia navigation system (INS) are as follows:

1. It is a totally global system. It enables an aircraft to fly great circle tracks and to navigate accurately across vast expanses of open sky where no ground base navigation aids are available, e.g., the North Atlantic or the Pacific Ocean.
2. It is a completely self-contained system and therefore is free from external navigation aids and atmospheric errors.
3. It is a very accurate system.

Note: Modern aircraft are fitted with two or more independent INSs, which allows them the advantage that their positions can be compared for possible system errors.

4. INSs that employ ring laser gyroscopes have the following advantages:
   a. Short warmup times
   b. No real wander
   c. No precession
   d. Extreme accuracy

What are the disadvantages of an INS?

The disadvantage of an inertia navigation system (INS) is it errors, which can fall into one of three categories:

1. *Bounded errors,* which are errors that do no keep increasing with time or increase and decrease in a cycle. For example:
   a. *Schuler loop.* Only common to a stable-platform INS that has been programmed to remain horizontal as the aircraft moves around the surface of the earth. The error exists at the first accelerometer level, i.e., acceleration, which can be passed up through the integration to affect velocity and distance, resulting in a distance error during the Schuler loop cycle, but the error returns to zero at the end of the cycle.
   b. *North alignment* error. This will produce bounded velocity errors.
2. *Unbounded errors,* which are errors that continue to increase as time goes on. For example:
   a. *Initial position* errors. Incorrect inputs into the system will create unbounded errors in velocity and position.
   b. *North alignment* error. This will produce serious unbounded errors in position.
3. *Inherent system errors.* For example:
   a. The INS position computer makes no allowance for a distance between two points being greater at height than on the surface because of the curvature of the earth. Fortunately, these differences are not large.
   b. The INS position computer also makes no allowance for the fact that the earth is not a true sphere.

As a result, accumulated errors will cause the INS position at the end of the flight to be different from the ramp position. The magnitude of this accumulated error is known as *radial error rate* and should be checked at the end of each flight to determine if the INS is operating within its defined limits of accuracy.

Describe GPS.

The Global Positioning System (GPS) is the U.S. Department of Defense (DoD) satellite system for worldwide navigation.

The GPS consists of 24 satellites, of which 21 are operational at any one time. This constellation of satellites is broken down into six orbital planes, each consisting of three or four satellites. Each circular orbital plane is at 55 degrees to the equator, with the satellites at a height of 20,200 km (10,900 nautical miles) and covers an orbit of the earth every 12 hours.

Four satellites will always be in line-of-sight range of an aircraft receiver at any position on the earth at any one time. The orbiting satellites transmit accurately timed radio signals, and the receiver equipment uses the time delay between transmission and reception to calculate its distance from the satellites. The distance measured from two satellites will establish a latitude and longitude fix. The distance measured from a third satellite will confirm the fix, and the fourth satellite will give altitude information.

Note: GLONASS is the Global Orbiting Navigation Satellite System operated by the Commonwealth of Independent States (CIS), formerly the Soviet Union, and is similar to the U.S. GPS.

What are the advantages of the GPS?

The advantages of the Global Positioning System (GPS) are as follows:

1. Truly global.
2. High-capacity use.
3. High-redundant satellite capacity system.
4. Built-in confirmation of the aircraft's position from the third satellite.
5. Cornerstone of future air navigation systems, including Global Landing Systems (GLS), which offers the advantage of a curved approach path and missed-approach guidance information at a lower installation cost compared with a normal instrument landing system (ILS) facility. The first GLS received Federal Aviation Administration (FAA) approval in 1997.
6. Ability to integrate GPS into other flight management systems.
7. Potential to be very accurate.
8. Ability to fly great circle tracks accurately.
9. Free (although rumors of imposed charges are often floated).

What are the disadvantages of the GPS?

The disadvantages of the Global Positioning System (GPS) for civilian use are twofold:

1. Downgrading of the system's accuracy, i.e., selective availability
2. System errors

1. *Selective availability.* The potential accuracy of the system is deliberately downgraded for civilian users by the U.S. Department of Defense for security reasons. This is known as *selective availability,*

but a better description is *selective unavailability.* However, as part of a 1996 presidential directive, President Clinton committed to discontinue selective availability in 2006, but, in fact, the White House decided to shut it off 6 years ahead of schedule on May 1, 2000. This move has allowed civilian GPS users to enjoy the same high accuracy as that enjoyed by the U.S. military. It also will help the FAA in its commitment to adopting a GPS-based sole-means navigation system for civil aviation in the United States.

2. *System errors:*
   a. Clock bias
   b. Satellite ephemeris
   c. Ionosphere propagation
   d. Instrument/receiver
   e. Satellite geometry
   f. Signal jamming

Therefore, the normal GPS is unsuitable for precision landing approaches while these poor accuracy levels are present. (*See Q: How does differential GPS work? page 96.*)

What is differential GPS?

Differential global positioning is a more accurate global positioning system (GPS) than a normal GPS. It applies a correction factor called *differential correction* to eliminate the two most significant errors in a normal GPS, namely, (1) selective availability (U.S. DoD downgrading of the normal GPS system for civil users) and (2) ionospheric errors.

How does differential GPS work?

Global Positioning System (GPS) signals are received at a ground installation, which has been surveyed accurately. The ground installation then computes the differences between the known position and the position at which the GPS says it is (known as *differential position error*). It then sends a differential correction factor to any aircraft within 70 nautical miles using an aircraft communications and reporting system (ACARS) link that enables the aircraft's onboard GPS navigation computer to correct its own normal GPS-derived position into a refined differential GPS position that is accurate to within 1 to 3 m.

Therefore, differential GPS has the potential accuracy to be suitable for precision landing approaches, i.e., Global Landing Systems (GLSs).

How is an INS/IRS better than GPS, especially for navigation information?

An inertia navigation system/inertia reference system (INS/IRS) is a better system than the Global Positioning System (GPS) for providing navigation information mainly because of the following:

1. The downgrading imposed on the normal GPS.
2. An INS/IRS is the only truly onboard self-contained system; therefore, it is not prone to external influences, i.e., natural effects or transmitter errors such as those which affect the GPS. (*See Q: What are the disadvantages of the GPS? page 95.*)

Therefore, the INS/IRS is a better navigation system because it is more self-contained and suffers fewer errors than the GPS.

What do you know about FANS?

Future air navigation systems (FANS) is the development of future navigation-based systems to meet the mutual demands of

1. U.S. and European airspace regulation changes. For example:
   a. Decommissioning of omega.
   b. Development of a free-flight airspace concept that incorporates the requirement for improved standards in
      (1) Navigation
      (2) Communication
      (3) Air traffic management (CNS/ATM)
      (4) Surveillance
2. The ever-increasing demand by the airlines for greater efficiency by reducing fuel consumption from more direct routes.

FANS is not a single specific system but the ongoing development of several navigation-based systems that meet the requirements of both the airspace regulators and the airlines. A first step toward future airspace systems is to improve en route navigation capabilities, including communications. This has seen the United States develop a satellite Global Positioning System, and Europe to favor a New Area Navigation System (*see Q: What is R Nav? page 98*) to be the basis of their new flight management systems. Both systems are fundamental parts of any air traffic management (ATM) system and are aimed at developing a free-flight airspace. (Free flight is an ATM system that allows pilots and airlines to set their own more direct flight paths away from the restrictions of airways.)

Free-flight plans also require the development of surveillance equipment that provides collision protection in the less regimented airspace.

The development of onboard traffic collision and avoidance systems (TCAS) and Automated Dependent Surveillance–Broadcast (ADS-B) systems satisfies this requirement for self-controlled collision protection. Therefore, TCAS and ADS-B can be categorized as a FANS. (*See Qs: What is TCAS? page 152; What is ADS-B? page 153.*)

What is R Nav?

R Nav is a form of onboard area navigation aircraft equipment that uses either a basic VOR/DME system or other position sensors, such as DME/DME, INS/IRS, Omega/VLF, or Loran C.

A simple area navigation system allows the operator to input the bearing and distance of a geographic location with reference to a given station (entered as VOR/DME bearing and range or an INS/IRS latitude and longitude) to position a waypoint, thus making the waypoint a specified geographic location. A series of waypoints is used to define an area navigation route or the flight path of the aircraft.

Note: A waypoint is sometimes likened to a phantom station because it provides the R Nav user with the same navigation information that a real VOR/DME installation would provide.

Thereafter, the pilot has access to deviation information from the track between waypoints by means of displayed information on the primary navigation instrument and distance to go from the DME reading.

The sophistication of the equipment determines the amount of information displayed and the ability of the system to automatically select and deselect the primary and alternative ground base navigation aids. The advantage of the system allows for more direct routing, i.e., not restricted to established airways, which is more efficient.

What do you know about the free-flight concept?

The free-flight concept is an air traffic management (ATM) system currently under development (since 1998, with trials in Alaska and Hawaii from 1999). It will allow pilots and airlines to set their own direct routes, altitudes, and speeds away from the restrictions of airways. The concept is based on two airspace zones, protected and alert. In principle, until the alert zone is breached, aircraft can be maneuvered with autonomous freedom, with the responsibility of collision avoidance in the alert zone moved to the pilot with the introduction of onboard TCAS and ADS-B systems.

The free-flight concept is the basis of the U.S. National Airspace System (NAS), which is planned to commence in the year 2005.

Ground Navigation Systems

What is decca?

Decca is a medium-range (300 to 500 nautical miles by day and 200 to 250 nautical miles by night) navigation system that uses ground-based beacons that operate in the low-frequency (LF) band using surface waves. It is capable of giving very accurate position information.

Decca uses hyperbolic position lines, which is a system that measures lines of equal difference in range between two beacons. Therefore, to fix your position, you need a minimum of three stations. The first two stations would determine the hyperbolic position line on which you are located, and the third station would give you a cross cut from a hyperbolic position line to obtain a minimum fix. Decca uses chains of four stations for added accuracy.

What is omega?

Omega is a long-range global navigation system that uses ground-based beacons and operates in the very low frequency (VLF, 3 to 30 kHz) band using surface-wave propagation.

What is rho rho rho omega?

Some omega systems use omega stations in a phase-measurement mode (such as VLF), which creates circular position lines with lanes a whole wavelength wide, i.e., 16 nautical miles. This principle is based on range and is called *rho rho* fixing for a two-station position fix or *rho rho rho* fixing for a three-station position fix.

What is Loran C?

Loran C is a long-range navigation system that uses ground-based beacons that operate in the low-frequency (LF) band using surface waves to achieve ranges of more than 100 nautical miles. It is available in North America, the North Atlantic, Europe, and the Mediterranean.

Describe VHF directional finding (VDF).

VHF directional-finding (VDF) uses communication radio waves and does not require any additional instrumentation in the cockpit.

It can be requested from air traffic control (ATC) at aerodromes that are equipped with radio aerials that can sense the direction of the very high frequency (VHF) com radio signals from an aircraft whenever the pilot transmits. This is known as *automatic direction finding* (DF) in the VHF band.

This system allows ATC to provide the pilot with aural bearing information on request, either as a QDM, QDR, QTE, or a QUJ.

Give the following Q definitions.

QUJ True bearing *to* the station

QTE True bearing *from* the station

QDR Magnetic bearing (radial) *from* the station

QDM Magnetic bearing (radial) *to* the station

What is a QGH letdown?

A QGH is a type of VHF directional-finding (VDF) letdown, usually available from military controllers.

The controller passes headings to steer and descent instructions to the pilot, similar to a radar letdown. The pilot does not have the responsibility of allowing for drift to bearings, etc., which is the case for a normal directional-finding letdown procedure.

A QGH letdown procedure has no approach chart, although minimum descent heights, etc., still apply.

Describe how an NDB/ADF system works.

A nondirectional beacon (NDB) is a medium-range radio navigational aid that sends out a signal in all directions for aircraft to home to. The NDB transmits in the 200- to 1750-kHz medium- and low-frequency bands and uses a surface-wave propagation path.

The automatic direction finder (ADF) is a needle indicator fitted in the aircraft that shows the direction to the selected NDB from the aircraft.

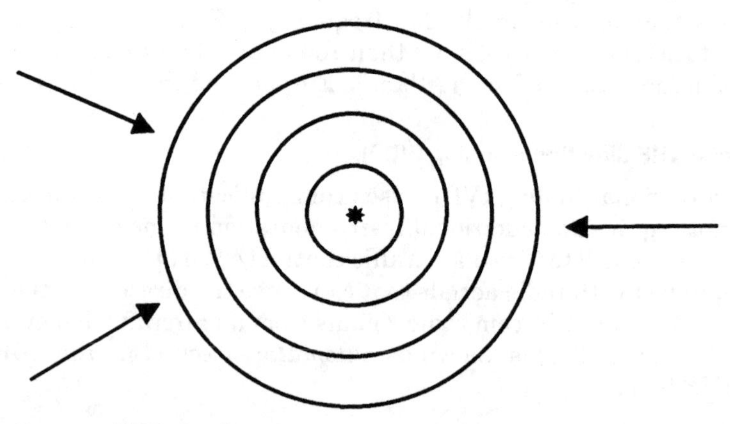

Figure 4.2 NDB signal.

This is either displayed as a relative bearing (angle between the aircraft heading and the direction of the NDB) on a relative bearing indicator (RBI) instrument or as a QDM on an RMI instrument.

The NDB/ADF system is used for en route tracking, holds, and NDB letdown procedures where low-powered NDBs are used as locators.

What is the range of an NDB?

The range of a beacon is controlled primarily by its power, with a maximum theoretical range of 300 nautical miles over land and 600 nautical miles over the sea, although its range is often restricted to avoid interference with other signals. For example:

$$\text{Range over land (in nautical miles)} = 3\sqrt{\text{power output (in watts)}}$$

$$\text{Range over sea (in nautical miles)} = 9\sqrt{\text{power output (in watts)}}$$

What are the errors of an NDB/ADF?

The errors suffered by an NDB/ADF are as follows:

1. Interference from other NDB stations
2. Static, especially thunderstorms
3. Night effect
4. Coastal refraction
5. Mountain effect
6. Aircraft quadrantal error
7. Synchronistic transmission

The International Civil Aviation Organization (ICAO) requirement for an NDB is that it maintains its accuracy to within ±5 degrees. However, not all errors are predictable, and therefore, it is possible that the actual error experienced by an NDB/ADF system is greater than its designated maximum.

What is an ADF beat frequency oscillator used for?

A beat frequency oscillator (BFO) is selected by the pilot on an automatic direction finder (ADF) navigation box and is used when identifying any nondirectional beacons (NDBs) that use unmodulated carrier waves, i.e., NON A1A transmissions. This is needed because no audio message (Morse) can be heard by the pilot on an unmodulated carrier wave, so the BFO imposes a tone onto the carrier wave to make it audible to the pilot; thus the NDB signal can be tuned and identified. However, it should not be left on when using the ADF to navigate.

Note: NDBs with A2A idents already modulate to an audible frequency. Therefore, the BFO is not needed and should always be switched off.

What is a VOR, and how does it work?

A VHF omni range (VOR) is a short-range sophisticated and accurate VHF navigational radio aid that outwardly generates specific track/position lines. A VOR ground transmitter radiates line-of-sight signals in all directions. However, unlike a nondirectional beacon (NDB), the signal in any particular direction differs slightly from its neighbor. These individual direction signals can be thought of as tracks or position lines radiating out from the VOR ground station. By convention, 360 different and separate tracks away from the VOR are used, each with its position related to magnetic north, i.e., 000 to 359 degrees.

Note: These 360 separate radials are achieved by the station transmitting 360 different VHF radio signals with both (1) an FM reference-phase signal, which is omnidirectional, i.e., the same in all directions, and (2) an AM variable-phase signal, which has its phase varying at a constant rate, thus creating unique signals for each of the 360 radials.

These VOR tracks or position lines are called *radials*; i.e., a radial is the magnetic bearing outbound from a VOR (QDR).

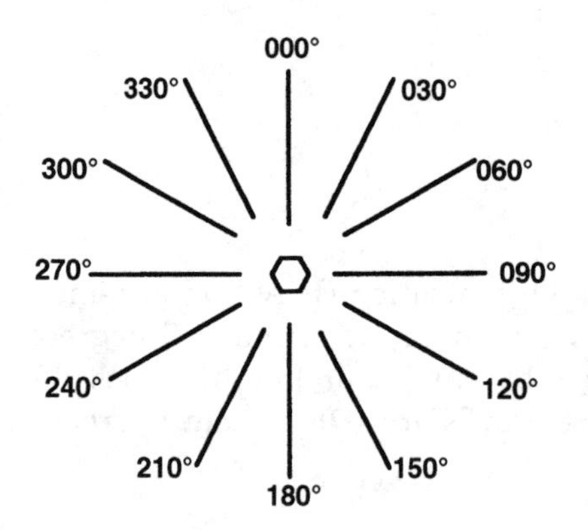

Figure 4.3 VOR signals—track/position lines.

The VOR transmits in the 108- to 118-MHz very high-frequency (VHF) band.

Note: The 108 to 112 MHz band is shared with instrument landing system (ILS) stations, so VORs are only allocated frequencies at even 100-kHz spacing, e.g., 108.2 kHz. Between 112 and 118 MHz, this band is used exclusively for VOR stations, and therefore, the frequency spacing is reduced to 50 kHz.

What is the range of a VOR?

The VHF omni range (VOR) uses VHF (radio signal) line-of-sight propagation paths. Its theoretical maximum range (nautical miles), based on line-of-sight propagation, is equal to

$$1.25\sqrt{H_1 \text{ (in feet AMSL)}} + 1.25\sqrt{H_2 \text{ (in feet AMSL)}}$$

where H_1 is the height of the transmitter and H_2 is the height of the receiver.

Note: However, its range is also a function of the power output of the transmitter.

What errors do a VOR experience?

The errors experienced by a VHF omni range (VOR) are as follows:

1. *Equipment errors* (both ground-based and airborne internal equipment errors). The total error of a standard airway VOR determines the distance between two VORs so that it maintains the airway's centerline within an airway's maximum permissible width, e.g., 5 nautical miles, with a maximum error of 5 degrees.
2. *Site error.* Site error is the effect of the signal being reflected off objects near to the beacon that causes tremendous confusion to the reading.
3. *Propagation errors.* Scalloping effect and atmospheric ducting.

What has the furthest range, a VOR or an NDB?

A nondirectional beacon (NDB) in theory has the farthest range because of its surface-wave propagation path, i.e., 600 nautical miles over sea and 300 nautical miles over land, compared with the VOR's line-of-sight propagation path. (*See Qs: What is the range of a VOR? page 103; What is the range of an NDB? page 101.*)

Describe an instrument landing system (ILS) and how it works.

The instrument landing system (ILS) is a precision approach radio aid that gives slope and track guidance to enable low-minima approaches for suitably equipped aircraft.

An ILS has two separate ground transmitters:

1. *The localizer.* The purpose of the localizer (azimuth) beam is to provide tracking guidance along the extended runway centerline, i.e., azimuth guidance left and right of the extended runway centerline. The localizer transmitter is usually in line with the centerline

at the end of the runway, and the signal is protected 10 degrees either side of the centerline up to a height of approximately 6000 ft and out to 25 nautical miles and 35 degrees on either side of the centerline out to 17 nautical miles.

2. *The glide slope.* The purpose of the glide-slope (elevation) beam is to provide vertical guidance toward the runway touchdown point, i.e., vertical guidance above and below the glide slope. The glide slope is normally set at an angle of 3 degrees to give a reasonable rate of descent glide path. The glide slope transmitting aerial is usually situated about 300 m in from the runway threshold to ensure adequate wheel clearance over the airfield fence. Therefore, when flying a glide slope, the aim is not to touch down on the runway's piano keys but the touchdown zone near to where the glide slope intersects the runway. The coverage of the glide slope signals extends to 8 degrees on either side of the localizer centerline out to 10 nautical miles.

The aircraft needs its own ILS receiver to decipher the localizer and glide-slope information and then to display it on either

1. A purpose-built ILS indicator (*see Q: Describe the purpose-built ILS indicator, page 145*).
2. A horizontal situation indicator (HSI) instrument (*see Q: Describe the HSI Instrument, page 146*).

Both are command instruments, and by making adjustments to maintain the localizer centerline and the glide-slope path, the aircraft will arrive at minimas exactly in line with the runway for landing (i.e., fly toward the deflected needles to center both the localizer and the glide slope as a center cross display).

The ILS localizer works in the 108- to 112-MHz VHF band, which it shares with the VOR. To avoid confusion with VOR signals, the ILS uses frequencies at odd 100- and 150-kHz spacing.

The ILS glide path uses the 329.3- to 335-MHz UHF frequencies at 150-kHz spacing. The glide path frequency is selected automatically when its paired VHF localizer channel is selected. Distance-measuring equipment (DME) is usually paired with the ILS frequency so that it is selected automatically with the ILS.

What is an ILS backcourse approach?

Some instrument landing system (ILS) installations provide a mirror image of the localizer beam on the reciprocal runway known as a *backcourse beam.* Like any mirror image, right is left and left is right, and an aircraft attempting an approach using a backcourse beam would display exactly the opposite of the correct indications unless the aircraft is fitted with a backcourse selector that reverses the signal internally to show the correct display.

Backcourse procedures can be found in the United States and Canada but are prohibited in the United Kingdom and most of Europe.

What errors does an ILS experience?

The errors that an instrument landing system (ILS) experiences are as follows:

1. *False glide slope.* The ground transmitter creates a mirror 150-Hz side lobe that overlaps the top of the 90-Hz main lobe to produce a false glide slope at approximately twice the angle of incidence above the real glide slope (e.g., 3 degrees real glide slope, 6 degrees false glide slope). By attacking the real glide slope from underneath, you ensure that you never encounter the false glide slope. However, if you were to follow the false glide slope, it should be recognized easily by the aircraft's high rate of descent, typically 1500 ft/min for a 6-degree glide slope.
2. *Phantom signals.* Outside a localizer's protected range, its signal cannot be relied on. This is so because the signal received is likely to be from another station using a similar frequency. This occurrence is rather common because of the small frequency range set aside for all the ILS stations.
3. *Backcourse approaches.* An aircraft attempting an approach using a backcourse beam would display exactly the opposite of the correct indications, especially on simple ILS or VOR indicators. However, the HSI display can be corrected by
   - *a.* Rotating the display 180 degrees to make the backcourse indications act in the correct sense.
   - *b.* Switching to backcourse on the aircraft with backcourse selectors. This system reverses the signal internally to show the correct indications.

What is MLS?

The microwave landing system (MLS) is a development of the instrument landing system (ILS) that uses two beams, one in azimuth and one in elevation like an ILS, but provides the ability to fly a curved or off-center localizer beam and/or a curved or varied glide slope angle.

Both beams use the same frequency in the SHF band because the beams share the frequency by time-multiplexing the signals that send out azimuth, elevation, and other data signals in a predetermined sequence.

How does MLS work?

The microwave landing system (MLS) has a narrow, fan-shaped azimuth beam for the localizer that sweeps to and fro between the limits of its coverage, namely, 40 degrees left and right of the centerline out to a maximum distance of 20 nautical miles.

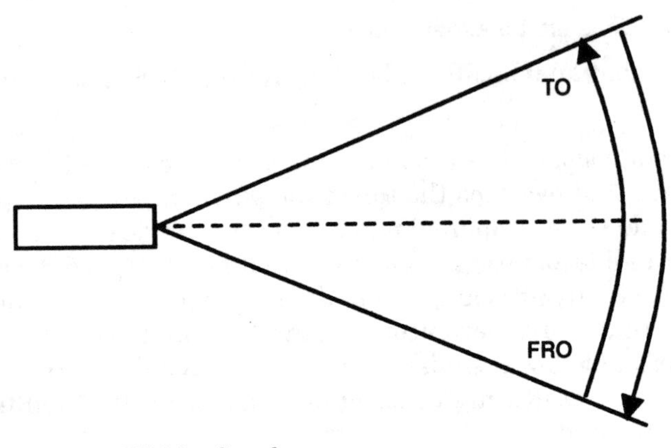

Figure 4.4 MLS localizer beam scan.

From the point of view of an aircraft on the approach, the azimuth beam starts on the left and sweeps to the right as the to scan and then back to the left as the fro scan. During the complete cycle of the to and fro scan, an aircraft would have the beam pass over it twice. An aircraft close to the left-hand side of the coverage would have a relatively large time interval between its to and fro sweeps passing over the aircraft. And an aircraft close to the right-hand side of the beam's coverage would have a relatively short interval. The time interval between the two scans is therefore a direct indication of the aircraft's position in the covered area. Similarly, the localizer path can be set to any line in the covered area in a similar manner. The elevation beam works in a similar way, but it has a horizontal, fan-shaped beam that sweeps from its lower limit of 0.9 degree to its upper limit of 15 degrees from the horizontal up to a maximum of 20,000 ft and out to 20 nautical miles within the coverage of the azimuth beam.

What are the advantages of the MLS?

The advantages of the microwave landing system (MLS) over the instrument landing system (ILS) are as follows:

1. MLS offers a varied-approach localizer (azimuth), either as a curved or off-center beam, and a glide slope (elevation) beam as either a curved or a varied glide-slope angle.

2. MLS beams are not sensitive to reflections from terrain and obstructions near the transmitter, which do affect ILS beams.
3. The MLS frequency range has more transmitting channels, i.e., 200 available channels, which remove the ILS problem of conflicting signals that occur because of the ILS's limited frequency range.

Describe the DME system and how it works.

Distance-measuring equipment (DME) is a form of secondary radar that gives continuous distance readout, in nautical miles, of the slant range to a ground station. DME operates in the ultrahigh frequency (UHF) band, between 962 and 1213 MHz, and uses PON transmissions with line-of-sight propagation paths.

DME consists of an onboard aircraft interrogator and a ground beacon transponder (which is opposite to secondary surveillance radar, where the ground equipment is the interrogator and the aircraft equipment is the transponder). The aircraft's interrogator initiates the exchange by transmitting a stream of pulses to the ground station, which then retransmits them back to the aircraft. The time delay between sending and receiving these pulses is converted into a range/distance.

Certain measures are built into the system to ensure that the transmitted and received pulses are correctly associated. These are as follows:

1. The aircraft identifies its own pulses, not those of other aircraft, because DME pulse trains are unique to each aircraft through the use of random pulse recurrence frequency (PRF; i.e., the number of pulses per second), also known as *jittering*. Thus the chances of two pulse trains being identical are effectively zero.
2. The aircraft also distinguishes between retransmitted pulses from the ground transponder and any reflected pulses off the ground surface because the ground transponder retransmits at a different frequency, namely, 63 MHz apart from the interrogator. In total, this system makes up 252 paired DME frequencies, which are called *channels,* in the UHF band.
3. The ground equipment will not be triggered by other non DME UHF transmissions because it will only reply to pairs of pulses separated by 12 μs, which is unique to aircraft DME interrogator signals.

The maximum range of a DME beacon is a function of the height of the aircraft in question and can be calculated by using the formula

$$\text{Range (nautical miles)}\sqrt{(1.5 \times \text{aircraft height in feet})}$$

The accuracy of all DME installations, as required by International Civil Aviation Organization (ICAO) has to be within ½ nautical mile or 3 percent, whichever is greater, in slant measurement.

The pilot generally selects the DME automatically when he or she selects a VOR or ILS frequency using the VHF NAV radio.

Note: Most DMEs are paired with either a VOR or a localizer frequency.

How is DME information used?

Distance/range information is an important navigational indication, especially when used in tandem with bearing information, such as the following:

1. A paired VOR/DME will give a very good en route position fix and/or distance to run to a waypoint (e.g., the radial bearing from the VOR and the DME distance along the radial).
2. Many ILSs have their localizer frequency paired with a DME beacon sited close to the runway threshold. This gives very useful terminal procedure information, namely:
   a. The distance to run to the runway threshold during an ILS or localizer approach.
   b. A DME arc procedure prior to intercepting the final approach track by maintaining an approximately constant distance reading while keeping another bearing aid (e.g., NDB) approximately 90 degrees abeam the aircraft nose.
3. By itself, DME range information will give a circle position line around a ground beacon.

What error does DME experience?

The only error of the distance-measuring equipment (DME) is that it measures a slant range and not a ground range; this is known as *slant-range error.* This is particularly evident when close to the beacon, i.e., when the ground distance is less than the height of the aircraft, and in this area, DME information should be disregarded. For example, when passing directly overhead a DME beacon, the DME indicator will show the height of the aircraft in nautical miles.

However, away from the beacon, the difference in slant and ground range is very small, and for most practical purposes, the DME distance can be considered to be correct, as the following example illustrates.
Formula:

$$\text{Ground range}^2 = \text{slant range}^2 - \text{aircraft height}^2$$

$$\text{Ground range}^2 = 50 \text{ nautical miles}^2 - 2 \text{ nautical miles}^2$$

Thus

$$\text{Ground range} = 49.96 \text{ nautical miles and slant range} = 50 \text{ nautical miles}$$

Additional operational misuse is a pilot error and cannot be considered a DME system error. For example, reading a DME display as a direct track to a beacon when the aircraft is flying abeam the beacon is a pilot error.

Radar

What does *radar* stand for?

The name *radar* was devised from *ra*dio *d*etection (direction) and *r*anging.

(*See Chapter 9, "Flight Operations and Technique," for definitions of radar information, advisory, and control services, page 279.*)

Describe primary (pulse) radar, and how it works.

Primary (pulse) radar works on the reflected signal principal. When transmitted electromagnetic radio energy encounters an object, it is reflected back to the point from which it was originally transmitted; therefore, it is used to detect aircraft. It does not require any specific equipment in the aircraft.

Primary radar uses ultrahigh and very high frequencies that have minimal static and atmospheric attenuation, and it uses a line-of-sight propagation path. Primary radar is also known as *pulse radar* and *reflective radar*.

What affects the range of primary pulse radar?

Atmospheric attenuation, antenna power, and the height of the aircraft all affect the range capabilities of a primary radar system.

1. *Atmospheric attenuation.* This reduces the range of a primary radar system. The greater the atmospheric attenuation, the more power is required to achieve the same range.
2. *Antenna power.* The greater the power output of the antenna, the greater is the range of the transmitted radio energy. To double the range, you have to quadruple the power.

3. *Height of the aircraft.* Because of the curvature of the earth, the higher the aircraft, the greater is the distance at which it can be detected by radar. That is,

$$\text{Radar range (nautical miles)} = \sqrt{1.5 \times \text{aircraft height in feet aboveground level (AGL)}}$$

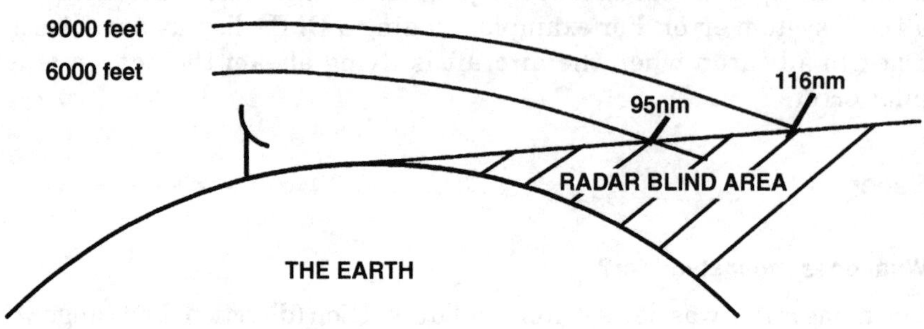

Figure 4.5 Radar range/aircraft height.

Describe secondary (surveillance) radar, and how it works.

Secondary radar works on the respondent-reply signal principle. It requires the active participation of an aircraft, which transmits a new return/reply signal back to the ground station whenever it receives an interrogation signal. This is accomplished by using an active onboard device called a *transponder*. The transponder device also allows the return signal to carry additional information, such as an identification code and even the aircraft's altitude, that allows the

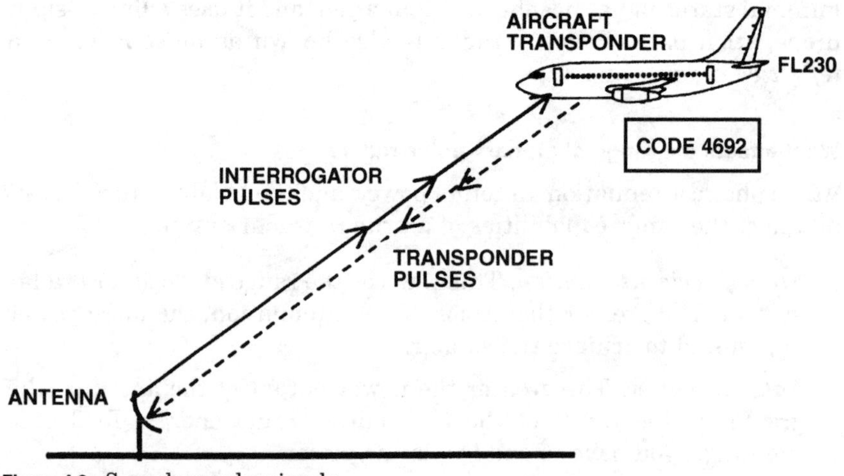

Figure 4.6 Secondary radar signals.

radar controller to easily identify a particular aircraft on his or her screen. Therefore, secondary radar is in fact two sets of radar talking to each other.

Describe the functions of a transponder.

The information carried on an airborne transponder's reply pulses is determined by the operating functions selected on the transponder unit by the pilot. The transponder facilities/functions are as follows:

1. *Code selection.* Knobs are provided for the pilot to select the aircraft's assigned identification transponder code. This four-number code is displayed in digital form on the transponder. The transponder should be selected to stand by (STDBY) while the code is being changed to avoid inadvertently transmitting an emergency or another aircraft's code.
2. *Transmission function switch.*

   STDBY. Usually the transponder is warmed up in the standby position prior to takeoff. From standby, the transponder is ready for immediate use.

   ON. This transmits the selected aircraft identification code in mode A (alpha) at the normal power level. (Mode A is the aircraft's identification mode only.)

   LOW SENS. Low sensitivity transmits the selected code at a lower power level than the ON position.

   ALT. Altitude, known as mode C (Charlie), transmits the aircraft's altitude in addition to the aircraft's normal selection code.

   TST. This selection causes the transponder to generate a self-interrogation signal to provide a check of its operation. If the transponder is operating correctly, its reply monitor light will illuminate.
3. *Reply / monitor light.* The reply monitor light on the transponder will flash to indicate that the transponder is replying to an interrogation pulse from a ground station.
4. *Ident switch.* This selection is made if air traffic control (ATC) requests a pilot to squawk ident. It causes a special symbol to appear for a few seconds on the radar screen, around the aircraft's return, thus providing a positive identification of the aircraft for the radar controller.
5. *System switch.* This switch selects transponder 1 or 2 when the aircraft is fitted with two transponders.
6. *Fault light.* This illuminates when the selected transponder fails.

What are the advantages of secondary radar?

The main advantage of secondary radar is that the reply pulse from the aircraft's transponder can carry coded additional information such as the following:

1. *Positive identification.* Air traffic control (ATC) uses secondary radar to distinguish a particular aircraft from others in its vicinity by assigning a special code, such as 1800 or 4200, which the pilot dials into his or her transponder, known as mode A (alpha).
2. *Abnormal situations.* Abnormal situations, such as radio failure, aircraft hijacking, and emergencies, are universally assigned special transponder codes that are used to communicate the aircraft's situation to ATC. These codes are dialed into the transponder by the pilot in place of any specific identity code.
3. *Altitude/ground speed reporting.* Most modern transponders are equipped with altitude and/or ground speed reporting, known as mode C (Charlie).
4. *Ident.* Identing an aircraft's responding signal provides a positive identification of the aircraft on a screen when the pilot presses the transponder's ident button. (*See Q: Describe the functions of a transponder, page 111.*)

For direction-measuring equipment (DME) questions, see subchapter, "Ground Navigation Systems," pages 107–109.

How does an isoecho radar work?

An isoecho radar, also known as a *weather radar,* is used primarily to detect thunderstorms and, by inference, severe turbulence. It is displayed on a black-and-white TV screen so that the aircraft can be navigated around the storms, if possible.

The system has a radar aerial fitted in the nose of the aircraft that can scan up to 90 degrees left and right and tilt 15 degrees up and down. It uses a pulse radar technique from a conical beam; pulse returns reflected back off water droplets in a cloud are interpreted and displayed by the system as black areas, and stronger returns that depict dense clouds and therefore high-intensity turbulence are displayed as hollow centers.

Note: Hollow centers are registered above a preset pulse return level. Because the equipment shows returns above a certain level differently (i.e., hollow centers), this system is known as isoecho circuitry.

Note. Color weather radar uses different colors to depict differing levels of returns and therefore different levels of storm activity and turbulence. (*See Q: On a color weather radar display, what colors represent the areas of greatest turbulence? page 113.*)

How do you calculate cloud tops (formula)?

Using the weather radar system, you can calculate the heights of cloud tops using the following formula:

$$\text{Height of cloud tops relative to the aircraft} = \frac{\text{Range (in feet)} \times (\text{weather radar scanner tilt} - \tfrac{1}{2}\ \text{beam width})}{60}$$

For example, for tilt 1 degree up, beam width 4 degrees and a cloud 50 nautical miles away,

$$\frac{50 \times 6080 \times (1 - 2)}{60} = -5067\ \text{ft}$$ (i.e., below the aircraft's cruising altitude)

What is AGC on weather radar?

AGC is automatic gain control.

AGC makes the system less sensitive to short-range returns so that it provides an even overall picture. Without AGC, you would expect the closer targets to produce a strong pulse return and therefore the closer targets would be brighter on the screen than the targets further away. AGC removes this misleading inaccuracy.

On a color weather radar display, what colors represent the areas of greatest turbulence?

Thunderstorms appear as red cores, and areas of maximum turbulence appear as magenta.

Note: The quantity of the radar pulses reflected back from a storm determines the severity of the storm.

Chapter

5

Atmosphere and Speed

Atmosphere

What is the aviation definition of height?

Height is the measured distance above the ground (i.e., QFE).

What is the aviation definition of altitude?

Altitude is the measured distance above the local pressure setting (i.e., QNH) or altitude above mean sea level (MSL).

What is the aviation definition of a flight level?

Flight level is the measured pressure level above the 29.92-in/1013-millibar datum.

What does ISA stand for?

International standard atmosphere. (*See Q: What are the ICAO ISA conditions at sea level? page 115.*)

What are the ICAO ISA conditions at sea level?

Temperature: +15°C [environmental lapserate (ELR), −2°C per 1000 ft)

Pressure: 29.92 inches of mercury or 1013 hectopascals (millibars) (decreases with altitude)

Density: 1225 g/m³ (density is influenced by temperature and pressure)

What is air pressure?

Air pressure is the weight of a column of air or the gravity force of air molecules. Air pressure is influenced by gravity and is proportional to density. Therefore, air pressure decreases with a rise in altitude.

The air pressure can be measured by using a simple mercury barometer, and the pressure experienced can be expressed as either (1) Inches of mercury (as used in the United States) or (2) force per unit area (e.g., millibars or hectopascals). These are both a unit of pressure force of a thousand dynes per square centimeter.

Note: Hectopascal is the standard unit of pressure measurement recognized by the International Civil Aviation Organization (ICAO).

How many feet are in a millibar (hectopascal)?

The number of feet in a millibar or hectopascal is variable because the rate that the pressure changes with height is variable. The air pressure (millibar) rate of change is more rapid near the earth's surface and less rapid at high altitudes.

By measuring the actual height of a single millibar for a given altitude, the rate of pressure change with altitude can be determined. The formula used to find the size of 1 millibar is

$$96 \times \frac{\text{absolute temperature (273K + actual temperature)}}{\text{altitude pressure (millibars)}}$$

For example, to find the size of 1 millibar at sea level given an altitude pressure of 1013 millibars and an outside air temperature (OAT) of +15°C, use

$$96 \times \frac{273 + 15}{1013} = 27.3 \text{ ft}$$

Therefore, at sea level/1013 millibars, 1 millibar = 27 ft; at 18,000 ft/500 millibars, 1 millibar = 40 ft; at 38,000 ft/200 millibars, 1 millibar = 80 ft. However, it is common to use 30 ft for 1 millibar up to an altitude of approximately 5000 ft.

How many inches of mercury to a millibar?

0.0295 inch of mercury equals 1 hectopascal/millibar.

What is pressure altitude?

Pressure altitude or pressure height is the international standard atmosphere (ISA) height above the 29.92-in/1013 millibar pressure datum, at which the pressure value experienced represents that of the level under consideration.

How do you calculate the pressure altitude's actual height?

To calculate a pressure altitude's actual height, you have to calculate the difference between the regional QNH and the 29.92-in/1013-millibar pressure datum. Then convert this pressure difference into a height, which you then add or subtract from the pressure altitude depending on whether the region QNH is above or below the 29.92-in/1013-millibar pressure datum. (Assume that 0.0295 inch of mercury/1 millibar = 30 ft unless otherwise instructed and that pressure decreases with height.)

Why do you need to calculate the actual height of a pressure altitude?

Any difference in the actual mean sea level (MSL) pressure from the international standard atmosphere (ISA) 29.92-in/1013-millibar pressure datum will give rise to a difference between the pressure altitude and the actual altitude. Therefore, when flying at a pressure altitude (flight level), it is important to know what your actual height is for the following reasons:

1. Ground clearance
2. Calculating the performance capabilities of the aircraft

Note: Traffic vertical separation should not be affected by a difference in pressure altitude and actual height because all aircraft should be on the same pressure subscale setting.

What pressure altitude error is commonly experienced?

Barometric pressure error on the altimeter.

Barometric pressure error is the effect of flying into an area with a different sea level pressure that gives rise to an incorrect altimeter reading. An aircraft normally would fly at a standard 1013-millibar/29.92-in pressure datum, and therefore, when it flies into an area with a lower actual sea level pressure, it has the effect of overreading the aircraft's altitude. Thus the aircraft is at a lower height than indicated—hence the adage "High to low beware below." Thus the opposite applies in the same manner.

What is OAT/SAT/TAT?

(*See Chapter 8, "Meteorology and Weather Recognition," page 224.*)

What is ISA temperature deviation?

International standard atmosphere (ISA) temperature deviation is the measurement of the actual temperature against the ISA temperature for the corresponding altitude. Temperature deviation is expressed as ISA ±6°C. In aviation, it is common practice to use ISA deviations as a means of describing temperature.

If you were at 33,000 ft with an outside air temperature (OAT) of −45°C, what is the temperature deviation from international standard atmosphere (IAS)?

ISA temperature is +15°C at mean sea level (MSL), with a 2°C per 1000 ft lapse rate. Therefore ISA temperature at 33,000 ft is

$$+15 + (33 \times -2) = -51°\mathrm{C}$$

Actual temperature minus ISA temperature at 33,000 ft is

$$-45°\mathrm{C} - (-51°\mathrm{C}) = +\ 6°\mathrm{C}$$

Therefore, temperature deviation at 33,000 ft = IAS + 6°C.

What is density?

Density is defined as mass per unit volume of a substance.

By definition, a block of gas is less dense than the same size block of liquid, which is less dense than the same size block of a solid.

What is density altitude?

Density altitude is the altitude above the 1013-millibar/29.92-in datum, at which the air density value experienced represents that of the level under consideration. That is, the altitude is measured against its air density value.

Density altitude allows for the consideration of international standard atmosphere (ISA) temperature deviation to pressure altitude.

What are the main influences on air density?

Air density is easily changed by compression and expansion due to the influences of

1. Temperature
2. Pressure
3. Humidity

Temperature is the main influence on air density for a given pressure altitude. That is, a warmer temperature equals less dense air.

Pressure also influences the density of a column of air over a range of altitudes. That is, the lower the altitude, the greater is the air density. The last influence on air density is moisture or humidity. Moist air is less dense than dry air because lighter water molecules replace some of the air's heavier molecules.

How does the variation of air density due to temperature affect an aircraft's performance?

A warmer than international standard atmosphere (ISA) temperature for a given altitude causes a decrease in air density, which causes a decrease in the performance level of the aircraft's engines and a decrease in aerodynamic performance because the aircraft's performance is based around air density. In addition, the opposite applies in the same manner.

What density errors are commonly experienced?

Altitude error and air-speed error.

Density for a particular pressure altitude is affected by temperature such that an altitude error will exist when the actual temperature experienced is different from the international standard atmosphere (ISA) temperature. This occurs because the aircraft's altimeter is basically only a barometer calibrated according to the ISA, which converts static pressure measurements to an altitude.

The altimeter does not make any allowance for temperature deviation from the ISA, and temperature deviation causes the density altitude to differ from the pressure altitude. This is so because colder air will make a parcel of air denser and therefore heavier, and this additional weight pushes down the column of air with the effect that pressure levels are experienced at lower altitudes than the ISA. This results in the actual flight level being lower than the pressure level read by the altimeter. In other words, the altimeter overreads. Therefore, when flying from high to low temperatures, beware below. This density altitude error is significant when considering terrain clearance.

The airspeed indicator (ASI) is calibrated to ISA mean sea level (MSL) conditions and measures dynamic pressure as the indicated airspeed (IAS). Because dynamic pressure is proportional, among other things, to air density, the ASI will only read the correct airspeed at one air density, namely, ISA MSL conditions of 1225 g/m^3. Any divergence from ISA MSL conditions will cause the ASI reading to be incorrect.

What is static pressure?

Static pressure is the equal weight of all the air molecules at any one point in the atmosphere.

Static pressure, as the name implies, does not involve any motion of the body relative to the air, but it is the ambient pressure around an object.

What is dynamic pressure?

Dynamic pressure is the pressure (over and above the static pressure) of the air molecules impacting onto a surface caused by either the movement of a body (e.g., an aircraft) through the air or the air flowing over a stationary object. Dynamic pressure is sometimes referred to as *moving pressure.*

There are two main components that determine the amount of dynamic pressure experienced:

1. The speed of the body relative to the air
2. The density of the air

These two components are used to formulate the actual measurement of dynamic pressure:

$$\text{Dynamic pressure} = \tfrac{1}{2}R\ (\text{rho}) \times V^2\ (\text{TAS})$$

where rho represents air density (which decreases with altitude and/or warmer temperatures), and V represents true air speed (the speed of the body relative to the air).

Speed

Describe IAS.

Indicated airspeed (IAS) is a measure of dynamic pressure translated to a speed and displayed to the pilot on the airspeed indicator (ASI) usually as knots per hour. Thus

$$\text{IAS} = \text{Dynamic pressure}\ (\tfrac{1}{2}RV^2)$$

However, because the design of the ASI instrument is calibrated to international standard atmosphere (ISA) mean sea level (MSL) conditions, it is only equal to the aircraft's true airspeed at ISA MSL conditions.

Describe RAS/CAS.

Calibrated airspeed (CAS) is the term commonly used in the United States, Australia, New Zealand, and many European countries, whereas *rectified airspeed* (RAS) is the term commonly used in the United

Kingdom. CAS/RAS is indicated airspeed (IAS) corrected for instrument and pressure errors.

Instrument error is caused by the inaccuracies in the construction of the airspeed indicator (ASI) itself, especially the friction of its moving parts.

Pressure error is also known as *position* or *configuration error.*

The calculated CAS/RAS figure is what the ASI would read if the particular ASI system were perfect.

Describe EAS.

Equivalent airspeed (EAS) is rectified airspeed (RAS) corrected for compressibility error. (*See Q: What are the effects of compressibility? page 24.*)

Describe TAS.

True airspeed (TAS) is simply the actual speed of an aircraft through the air mass in which it is flying.

TAS therefore depends on the air density properties of the air mass in which the aircraft is flying. The higher the air density, the greater is the resistance to the motion of the aircraft, and therefore, the lower is the TAS. Conversely, the lower the air density, the less is the resistance to the motion of the aircraft, and therefore, the higher is the TAS.

Air density varies for two main reasons:

1. *Temperature.* Warm air is less dense than colder air.
2. *Pressure.* The higher the altitude, the lower is the air density, and the lower the altitude, the greater is the air density. (*See Q: What are the main influences on air density? page 118.*)

Therefore, at a constant IAS/RAS, the TAS will increase with an increase in altitude or temperature because of the reduction in air density. This can be expressed by using the dynamic pressure/IAS formula:

$$\text{Dynamic pressure (IAS)} = \tfrac{1}{2}R \times V^2 \text{ (TAS)}$$

(*See Q: What density errors are commonly experienced? page 119.*)

Describe ground speed.

Ground speed is the actual speed of an aircraft relative to the ground.

Ground speed is a combination of the following:

1. True airspeed (the speed of an aircraft through an air mass)

2. Wind velocity (wind velocity represents the movement of an air mass relative to the ground)

Therefore, ground speed can be either greater than true airspeed (TAS) if the aircraft has a tailwind, less than the TAS if the aircraft has a headwind, or equal to TAS in still air conditions. Ground speed is used for navigation and flight-planning purposes.

Describe LSS.

LSS is the local speed of sound.

The speed of sound, Mach 1.0, does not remain a constant value of knots per hour because it is influenced by temperature; therefore, the speed of sound decreases with a decrease in temperature. The speed of sound for a particular temperature is known as the *local speed of sound.*

The local speed of sound can be found by using the following formula:

$$\text{LSS} = 38.94\sqrt{\text{absolute temperature}}$$

At international standard atmosphere (ISA) mean sea level (MSL) conditions, LSS is equal to 660 knots per hour, and we know that temperature decreases with an increase in altitude. Therefore, the LSS decreases with an increase in altitude.

How does temperature affect the LSS?

As temperature decreases, the local speed of sound (LSS) decreases.

$$\text{LSS} = 38.94\sqrt{\text{absolute temperature (i.e., 273 + actual temperature)}}$$

Describe Mach number.

Mach number is the true airspeed (TAS) of an aircraft, given as a percentage relative to the local speed of sound (LSS) and displayed to the pilot on the Mach meter instrument. For example, half the LSS would be shown as 0.5 Mach.

$$\text{Mach number (MN)} = \frac{\text{TAS}}{\text{LSS}}$$

Therefore, MN is a function of the behavior of TAS and the LSS. With an increase in altitude at a constant indicated airspeed (IAS), the MN increases because TAS increases with altitude, due to a decrease in air density, and the LSS decreases, due to the drop in air temperature.

Mach number (MN) is used as a speed reference at high altitudes, usually above 26,000 ft, because the MN becomes the aircraft's limiting speed in preference to IAS. That is, up to approximately 26,000 ft, an aircraft will climb at a constant IAS against an increasing MN, where the MN speed reaches the aircraft's MN limiting speed. Then above 26,000 ft the aircraft is flown at a constant MN with a decreasing IAS for an increase in altitude.

How do you calculate Mach number?

Mach number (MN) is a calculation of the true airspeed (TAS) relative to the local speed of sound (LSS):

$$\text{MN} = \frac{\text{TAS}}{\text{LSS}}$$

What is the main influence on Mach number?

Temperature.

As temperature decreases, the local speed of sound (LSS) decreases.

$$\text{LSS} = \frac{38.94}{\text{absolute temperature}}$$

If the LSS decreases for a constant true air speed (TAS), then the Mach number must rise.

$$\text{MN} = \frac{\text{TAS}}{\text{LSS}}$$

Mach number is a function of the LSS, which is influenced by temperature; therefore, Mach number is influenced by temperature.

What happens to the indicated Mach number (MN) in a long-range cruise as weight decreases at the same flight level (FL)?

It decreases.

The RAS/TAS/MN questions are too numerous to include all the possible variations. However, if you appreciate the effect of temperature and density and the basic relationships between the different speed measurements, then you should be able to answer any questions in this area. Beware of questions with inversion layers, where the temperature increases with altitude, or isothermal layers, where the temperature remains constant with a change in altitude. Here are some RAS(IAS)/TAS/MN question examples.

What happens to your calculated airspeed (CAS) or indicated airspeed (IAS) if you descend at a constant true airspeed (TAS)?

The CAS (IAS) will increase.

This is so because pressure increases with a decrease in altitude due to the gravity effect. Density is proportional to pressure. Therefore, density increases with a decrease in altitude, which means the CAS (IAS) increases in a descent at a constant TAS.

Alternatively, this can be expressed as

$$\text{CAS (IAS)} = \tfrac{1}{2}R \times V^2$$

where R (density) increases with a decrease in altitude, and V (TAS) is constant. Therefore, CAS increases. (*See Q: Describe IAS, page 120.*)

What happens to your true airspeed (TAS) if you climb at a constant calibrated airspeed (CAS) or indicated airspeed (IAS)?

The TAS will increase.

This is so because air density decreases with altitude, and this can be expressed as

$$\text{CAS (IAS)} = \tfrac{1}{2}R \times V^2$$

where CAS (IAS) is constant, and R (density) decreases with an increase in altitude. Therefore, V (TAS) will increase. (*See Q: Describe TAS, page 121.*)

What happens to your Mach number (MN) if you climb at a constant calibrated airspeed (CAS) or indicated airspeed (IAS)?

The MN will increase.

This is so because true airspeed (TAS) increases because air density decreases with altitude and the local speed of sound (LSS) decreases because of a reduction in air temperature with increases in altitude. Therefore, given that MN is a function of TAS divided by LSS, the MN will increase during a constant CAS (IAS) climb. (*See Q: Describe Mach number, page 122.*)

What happens to your calibrated airspeed (CAS) or indicated airspeed (IAS) and true airspeed (TAS) if you fly at a constant CAS (IAS) into a warmer area?

The CAS (IAS) remains constant because CAS (IAS) is unaffected by temperature, whereas the TAS will increase because of the decrease in air density associated with warmer temperatures.

If two aircraft are flying at different flight levels at the same Mach number, which aircraft has the higher TAS/CAS (IAS)?

The aircraft at the lower altitude has a higher true airspeed (TAS)/calibrated airspeed (CAS) or indicated airspeed (IAS).

This is so because the local speed of sound (LSS) decreases as temperature decreases with an increase in altitude.

$$\text{Mach number (MN)} = \frac{\text{TAS}}{\text{LSS}}$$

Therefore, to maintain the same MN at a higher altitude, the TAS must be reduced, and if the TAS is reduced, the CAS (IAS) will be reduced. Therefore, the aircraft at the lower altitude has the higher TAS/CAS (IAS).

What happens to Mach number (MN) and calibrated airspeed (CAS) or indicated airspeed (IAS) when climbing through an isothermal layer at a constant true airspeed (TAS)?

Mach number remains constant, and CAS (IAS) decreases.

The Mach number remains constant because an isothermal layer has a constant air temperature, and therefore, the local speed of sound (LSS) is constant. Thus, since

$$\text{MN} = \frac{\text{TAS}}{\text{LSS}}$$

and LSS and TAS are constant, the MN also must remain the same value.

The CAS (IAS) decreases because pressure decreases with altitude due to the gravity effect. Therefore, density decreases, which means that the rectified airspeed (RAS) decreases. This can be expressed as

$$\text{CAS (IAS)} = \tfrac{1}{2}R \times V^2$$

where V (TAS) is constant, and R (density) decreases with an increase in altitude. Therefore, CAS (IAS) decreases.

What happens to calibrated airspeed (CAS) or indicated airspeed (IAS) if an aircraft descends through an isothermal layer at a constant true airspeed (TAS)?

CAS (IAS) increases.

This is so because pressure increases with a decrease in altitude due the gravity effect. Therefore, density increases, which means rectified airspeed (RAS or IAS) increases. This can be expressed as

$$\text{CAS (IAS)} = \tfrac{1}{2}R \times V^2$$

where V (TAS) is constant, and R (density) increases with a decrease in altitude due to increased pressure. Therefore, CAS (IAS) increases.

What happens to the calibrated airspeed (CAS) or indicated airspeed (IAS) and Mach number (MN) if an aircraft climbs at a constant true airspeed (TAS) through an inversion layer?

CAS (IAS) decreases more than normal, and MN decreases.

CAS (IAS) decreases more than normal because an inversion layer has an increase in temperature with altitude. Therefore, the air density decreases more than normal.

$$\text{CAS (IAS)} = \tfrac{1}{2}R \times V^2 \text{ (TAS)}$$

Therefore, R decreases more than normal, which results in CAS (IAS) decreasing more than normal when climbing at a constant TAS.

MN decreases because the local speed of sound (LSS) increases due to the increase in temperature of the inversion layer and TAS is constant. Thus MN will decrease.

For performance speed questions, see Chapter 7, "Performance and Flight Planning," page 183.

Chapter

6

Aircraft Instruments and Systems

Pressure Flight Instruments

What are the pressure flight instruments, and how do they work?

The pressure flight instruments are

1. Airspeed indicator (ASI)/Mach meter (MM)
2. Altimeter
3. Vertical speed indicator (VSI)

Pressure instruments measure atmospheric pressure by using the pitot-static system, which is a combined sensor system that detects the following:

1. The total pressure (static and dynamic pressure), also called *pitot pressure,* which is measured by a pitot probe
2. Static pressure alone, which is measured by either the static port on a pitot probe or by a separate static vent

The difference between the two will give a measurement of the dynamic pressure. That is,

$$\text{Dynamic pressure} = \text{total pressure} - \text{static pressure}$$

Dynamic and/or static pressure measurements are the basis of the flight instrument readings.

How does the airspeed indicator (ASI) work?

The ASI measures dynamic pressure as the difference between the total pitot pressure measured in the instrument's capsule/diaphragm and the static pressure measured in the case. The dynamic pressure represents the indicated airspeed (IAS) as knots per hour. (*See Q: Describe indicated airspeed (IAS), page 120.*)

The ASI instrument is calibrated to international standard atmosphere (ISA) mean sea level (MSL) density of 1225 g/m^3.

What are the airspeed indicator (ASI) instrument errors?

The ASI instrument suffers from the following errors:

1. Instrument error
2. Pressure error
3. Density error
4. Compressibility error
5. Maneuver error
6. Blocked pitot static system

How is VMO displayed on the airspeed indicator (ASI)?

On the ASI display, a red/black striped pointer indicates the VMO speed.

Describe how a Mach meter works.

The Mach meter measures the airspeed relative to the speed of sound. In essence, the Mach meter is a combined airspeed indicator (ASI) and altimeter that comprises the following:

1. A capsule feed with pitot pressure inside an ambient static pressure feed case that acts as an ASI and measures dynamic pressure as the airspeed.
2. A sealed capsule containing international standard atmosphere (ISA) mean sea level (MSL) conditions inside the ambient static pressure feed case, which acts as an altimeter by measuring the static pressure, which it relates to an altitude.

$$\text{Mach meter ratio} = \frac{\text{pitot} - \text{static}}{\text{static}}$$

These two functions inside the Mach meter are linked via a ratio arm that itself acts on a ranging arm to ultimately move the pointer/digital display on the instrument face. Although the local speed of sound (LSS) is dependent only on temperature, no temperature sensors are found in the Mach meter because it uses the mathematically proven fact that the

$$\text{Mach number} = \frac{\text{dynamic pressure}}{\text{static pressure}}$$

Mach number is usually displayed in a digital form in a window on the ASI face; therefore, MN and the corresponding IAS are assessed easily.

What errors does a Mach meter suffer from?

The Mach meter only suffers from the following errors:

1. Instrument error, which is caused by the inaccuracies in the Mach meter's construction
2. Pressure error, also known as *position* or *configuration error*

However, these errors are extremely small, and therefore, the indicated Mach number speed can be read as the true Mach number speed.

3. Blocked pitot static system (*see Q: What are the airspeed indicator (ASI)/Mach number (MM) indications and actions for a blocked pitot static probe? page 129*).

Note: The Mach meter does not suffer from density or temperature errors because its built-in altitude capsule design and its ratio to the dynamic pressure measured compensate for density and temperature variations.

What are the airspeed indicator (ASI)/Mach meter indications and actions for a blocked pitot and/or static probe?

A static line blockage means that the static pressure in the ASI/Mach meter instrument case remains a constant value. Therefore,

1. At a constant altitude, the ASI/MM will read correctly.
2. During descent, the ASI/MM will *overread* due to an increase in the pitot pressure in the capsule against the trapped low static pressure of the higher altitude.
3. During climb, the ASI/MM will *underread* due to a decrease in the pitot pressure in the capsule against the trapped high static pressure of the low altitude.

A pitot line blockage means that the total pressure in the ASI/MM instrument capsule remains a constant value. Therefore,

1. At a constant altitude, the ASI/MM reading will not change even if the airspeed does due to the trapped pitot pressure in the capsule against a constant altitude static pressure.
2. During descent, the ASI/MM will *underread* due to an increase in the static pressure in the case against a constant pitot pressure.
3. During climb, the ASI/MM will *overread* due to a decrease in the static pressure in the case against a constant pitot pressure.

The actions for a blocked pitot/static system causing an unreliable ASI/MM reading would be to

1. Ensure that the pitot static probe anti-ice heating (pitot heat) is on, if applicable.
2. Use an alternative, such as a static source or an air data computer, if applicable.
3. Use a limited flight panel, i.e., standby ASI.
4. Fly at a correct attitude and power setting.

How does a pressure altimeter work?

A simple pressure altimeter is designed to measure static air pressure, which it relates to an indicated altitude. As the aircraft ascends, the static pressure in the instrument case decreases, which allows the enclosed capsule to expand, and this in turn moves the needle on the instrument face to indicate a corresponding altitude. For a descent, the opposite function applies.

A subscale setting device is included so that the instrument can be zeroed to various datum elevations. (*See Q: Give the definitions of the following altimeter subscale settings, page 130.*)

The altimeter capsule is calibrated to full international standard atmosphere (ISA) mean sea level (MSL) conditions, i.e., +15°C, 29.92 in/1013 millibars, and 1225 g/m^3.

Give the definitions of the following altimeter subscale settings?

QNH. QNH is a local altimeter setting that makes the altimeter indicate the aircraft's altitude above mean sea level (AMSL) and therefore airfield elevation. There are two types of QNH:

1. Airfield QNH
2. Regional QNH, which is the lowest forecast QNH in an altimeter setting region. QNH is QFE reduced to sea level using international standard atmosphere (ISA) values for the calculation.

QFE. This zeros the altimeter on the airfield elevation datum. There are two types of QFE:

1. Airfield QFE is measured at the highest point on the airfield.
2. Touchdown QFE is measured at the touchdown point of the runway in use for precision approaches.

QFF. This is similar to QNH except that it uses the actual conditions (not ISA) to find the sea level pressure. It is used more commonly by meteorologists than by pilots.

QNE. This is not an altimeter setting but is the height shown at touchdown on the altimeter with 29.92 in or 1013 millibars (hPa) set on the subscale. It is used at very high aerodromes where QFE pressure is so low that it cannot be set on the altimeter subscale.

Standard setting. 29.92 in or 1013 hPa millibars standard setting will give altimeter readings as a pressure altitude or flight level and is used for traffic controlled airspace above the transition layer.

What are the aviation definitions of height, altitude, and flight level?

(*See Chapter 5, "Atmosphere and Speed," page 115.*)

What are the altimeter instruments errors?

The altimeter instrument errors are as follows:

1. Instrument error
2. Pressure error (also known as *position* or *configuration error*)
3. Time-lag error
4. Barometric error
5. Temperature/density error
6. Blocked static port

What are the altimeter indications and actions for a blocked static port?

A static line blockage means that the static pressure in the altimeter instrument case remains a constant value. Therefore, the altimeter will display the altitude where the blockage occurred regardless of any actual change in the aircraft's altitude.

The actions for a blocked static line causing an unreliable altimeter reading would be

1. To ensure that the pitot static probe anti-ice heat is on (pitot heat), if applicable.
2. To use an alternative, such as a static source or an air data computer, if applicable.

3. To use a limited flight panel, i.e., standby altimeter or vertical speed indicator (VSI), if available.
4. To fly correct attitude and power settings, especially for level flight.

For climbs and descents, calculate rate of climb or descent (ROC/ROD) either from VSI or from approximates given the aircraft's attitude and power performance against time from known altitudes to determine present altitude.

Given a temperature deviation from ISA of 36°C, the pressure altimeter will (a) overread, (b) underread, or (c) read correctly, and why?

The altimeter will overread because the temperature deviation is colder than the international standard atmosphere (ISA); i.e., the altimeter reads an altitude higher than the actual altitude of the aircraft. (*See Q: What density errors are commonly experienced? page 119.*)

What do you know about servo-assisted altimeters?

A servo-assisted altimeter increases the accuracy of a simple pressure altimeter because its design no longer relies on a direct mechanical link between its capsule and the altitude pointer on the instruments display dial. Instead, the servo-assisted altimeters use an electrically conducted E&I bar arrangement.

What is the advantage of a servo-assisted altimeter?

The advantage of a servo-assisted altimeter is its improved design feature between the capsule and the display pointer that removes (1) instrument error and (2) time-lag error, which are associated with a simple altimeter. (*See Q: What are the altimeter instrument errors? page 131.*)

How does a vertical speed indicator (VSI) instrument work?

The VSI instrument measures the rate of change of static pressure and displays this as a rate of climb or descent (expressed as feet per minute, or fpm) on the VSI instrument face. The capsule is fed with static pressure and reacts immediately to any change in static pressure, whereas the static pressure feed into the case is restricted or slowed by a metering unit, thus creating a differential static pressure between the capsule and the case. As long as the aircraft continues to climb or descend, the VSI will translate this as a rate of climb or descent measurement on the instrument dial face.

What errors do the vertical speed indicator (VSI) instrument suffer from?

The errors that the VSI instrument suffers from are

1. Instrument time-lag error
2. Pressure error (also known as *position error*)
3. Maneuver error

What type of dial display does a vertical speed indicator (VSI) have?

The VSI uses a logarithmic scale display that has a greater sensitivity at small rate of climb or descent (ROC/ROD) values. Zero is usually at the 9 o'clock position with an ROC scale above and an ROD scale below.

What do you know about an instantaneous vertical speed indicator (IVSI)?

The IVSI was designed to counter the time-lag error experienced by simple VSI. The IVSI uses two spring-loaded dashpots in the static line before the capsule that cause an immediate differential pressure to be sensed due to their inertia at the start of a climb or descent. Once the aircraft is established in a climb or descent, the dashpots are centered by their springs, and when the aircraft starts to level out, the opposite inertia of the dashpots produces an immediate change in the reading on the IVSI display.

What are the advantages of an instantaneous vertical speed indicator (IVSI)?

The advantage of an IVSI is the immediate display of any change in the aircraft's rate of climb or descent (ROC/ROD).

What are the disadvantages of an instantaneous vertical speed indicator (IVSI)?

The disadvantage of an IVSI is that the dashpots, which sense the vertical acceleration of the aircraft, are also affected by the acceleration in a turn. Therefore, the IVSI has an error that it initially shows as a rate of climb (ROC) when applying large angles of bank, i.e., over 40 degrees of bank. However, if the turn is maintained, the IVSI will stabilize to zero but then indicates a rate of descent (ROD) as the aircraft rolls out of the turn.

What are the vertical speed indicator (VSI) indications and actions for a blocked static port?

A static line blockage means that the static pressure in the VSI instrument capsule and case via the metering unit both remain a constant value. Therefore, the VSI display will read zero at all times regardless of any actual change in the aircraft's rate of climb of descent (ROC/ROD).

The actions for a blocked static line causing an unreliable VSI reading would be

1. To ensure that the pitot static probe anti-icing heating (pitot heat) is on, if applicable.
2. To use an alternative, such as a static source or an air data computer, if applicable.
3. To use a limited flight panel, i.e., altimeter, if available.
4. To fly correct attitude and power settings.

For climbs and descents, calculate rate of climb or descent (ROC/ROD) either from the altimeter or from approximates given known aircraft attitude and power performance.

How is air temperature measured?

Either a total head thermometer or a rosemount probe that is extended into the free airstream commonly measures air temperature. Usually this temperature is displayed to the pilot on a total air temperature (TAT) gauge. (*See Q: Describe total air temperature, page 224.*)

In flight, TAT is only a function of the ram effect of the air entering the probe, and pitot heat is not considered when calculating outside air temperature (OAT).

What do you know about air data computers?

Modern aircraft feed their static and pitot lines into an air data computer (ADC) that calculates the RAS, TAS, MN, TAT, ROC, and ROD and then passes the relevant information electronically to the servo-driven flight instruments (not the standby instruments, which retain their own direct static/pitot feeds).

The advantage of the ADC system is that the data calculated can be feed to the following:

1. Autopilot (AP)
2. Flight director system (FDS)
3. Flight management system (FMS)
4. Ground proximity warning system (GPWS)
5. Navigation aids
6. Instrument comparison systems

Gyroscopic Flight Instruments

What are the gyro flight instruments?

The gyroscopic flight instruments are

1. Directional indicator (DI)
2. Artificial horizon (AH)
3. Turn and slip indicator or turn coordinator

What is a gyroscope?

A gyroscope is a body (usually a rotor/wheel) rotating freely in one or more directions that possesses the gyroscopic properties of rigidity and precession.

How does a gyroscope work?

A gyroscope measures the force experienced on its rotor body during a maneuver of an aircraft. The rotor is usually suspended in a system of frames called *gimbals,* which are arranged at right angles to each other and are used as a conduit to transfer the force experienced on the rotor to a displayed measurement on the instrument face.

The gyro's rigidity property allows it to remain stable in space while the aircraft moves around it. Any force applied to the gyro as a result of the aircraft changing direction or attitude around the gyro will precess the gyro. This precession is translated into a movement of the gimbals that are attached to a direction scale (on the directional indicators) or an attitude horizon bar (on the artificial horizon) that measures and displays the movement of the aircraft on the instrument face. One gimbal is required for each axis around which you wish to measure a precessed movement.

The spin axis of the gyroscope and the axis of precessed movement measured by the gimbal represent a gyro's planes of freedom, and a gyroscope is classified in terms of its planes of freedom. In total, a gyroscope can have three planes of freedom: the spin axis of the gyroscope and two other possible planes at right angles to the gyroscope's spin axis and each other.

A gyroscope's rigidity is all-important and is itself a product of the spin (drive) speed of the rotor, which can be either vacuum (air) or electrically driven.

What is gyroscopic wander?

Any movement of the gyroscope's spin axis away from its fixed direction is called *wander,* and if this occurs, it gives rise to inaccurate instrument readings.

The rigidity of a gyroscope system is responsible for maintaining the direction of the spin axis and is fixed in space.

What is the gyroscope caging system, and why is it used?

A caging system locks the gyroscope, i.e., artificial horizon, in a fixed position, especially when the gyroscope is not being used, i.e., when parked, and for some aircraft, it is recommended during aerobatic maneuvers. Caging a gyroscope in this manner will prevent it from toppling (rigid in space), and thus when the aircraft is restarted, the instrument reaches its fully erect position very quickly.

What is real wander of a gyroscope?

Real wander occurs whenever the direction of a gyroscope's spin axis actually moves from its alignment in space. If this occurs, it gives rise to inaccurate instrument readings. Real wander can be either (1) induced deliberately by applying an external correcting force, e.g., as in the alignment of tied gyroscopes, or (2) caused by imperfections in the gyroscope, e.g., unbalanced gimbals or bearing friction.

Note: This is also known as mechanical drift.

Describe the directional indicator (DI) instrument.

The DI is a gyroscope that displays the aircraft's heading using a compass rose display.

The DI consists of

1. A tied gyroscope.
2. The gyroscope rotates about the earth's horizontal axis.
3. Two gimbals.
4. Three planes of freedom (i.e., the gyroscope's spin axis and the pitch and roll axes of the gimbals).
5. The gyrocope axis is aligned (direction) to true north.

What is apparent wander of a gyroscope?

Apparent wander is a natural phenomenon. A directional indicator (DI) gyro appears to wander not because of any real changes in the direction of the gyroscope's spin axis alignment in space but because its orientation has changed due to rotation of the earth. If this occurs, it gives rise to an incorrect heading display. (*See Q: What is gyroscopic wander? page 135.*)

How do you correct for apparent wander?

Depending on the sophistication of the instrument, apparent wander is corrected in one of two main ways:

1. By periodically (10- to 15-minute intervals) realigning the directional indicator (DI) to the magnetic compass heading using the slaving knob on the DI.
2. If the DI instrument is fitted with a latitude nut, it produces an opposite error to the earth's rotation (i.e., 15 × sin latitude in degrees per hour) to give an adjusted heading. If, however, the aircraft was moved away from the latitude at which the lat nut was set, then an error, either positive or negative, would arise because the degree of apparent wander varies with latitude; i.e., it increases toward the poles.

Note: On early instruments, this was a real nut that was screwed in or out to provide the necessary imbalance in drift on the gyroscope's spin axis for the aircraft's actual latitude. On modern aircraft, this effect is accomplished by onboard computer software adjustments.

What is transport wander on an uncorrected gyroscope?

Transport wander is a form of apparent wander. If a directional indicator (DI) gyroscope is aligned to true north at one place on the earth and then the aircraft is moved to another east-west position on the globe, the gyroscope axis will be out of alignment. This is known as *transport wander* on uncorrected gyroscopes.

Transport wander + apparent wander = total apparent wander

Flights north-south produce no transport wander but will produce apparent wander as the latitude changes.

What errors do a directional indicator (DI) suffer?

A DI suffers from (1) gyroscope system failures and (2) total wander errors.

What is the advantage of the directional indicator (DI) over the magnetic compass?

The rigidity of the DI gyroscope gives steadier heading information than the magnetic compass, which suffers from turning and acceleration errors.

Describe the air-driven artificial horizon instrument.

The artificial horizon is the primary attitude instrument that measures and displays the pitch and roll of the aircraft about the horizon level.

The artificial horizon consists of the following:

1. An earth gyroscope.
2. The gyroscope rotates about a vertical axis.
3. Two gimbals.
4. Three planes of freedom
   a. The gyroscope's spin axis.
   b. Pitch and roll axis of the gimbals.
5. The gyroscope's axis is aligned to the earth's vertical.

Simply, the aircraft moves around the artificial horizon gyroscope, and the gimbals measure the aircraft's pitch and roll maneuvers.

What errors do an artificial horizon experience?

The artificial horizon experiences the following errors:

1. Turning errors
2. Acceleration errors
3. Real wander of the gyroscope's spin axis away from its alignment with the earth's vertical. (*See Q: What is real wander of a gyroscope? page 136.*)

Turning and acceleration errors in the artificial horizon gyroscope are caused by lateral acceleration in turns and by the aircraft's acceleration and deceleration forces that induce a false position indication of the gyroscope's vertical axis.

Note: Remember, the artificial horizon is an earth gyroscope with its spin axis in the earth vertical.

What are the indications and actions for a failed artificial horizon?

The indications of a failed air-driven artificial horizon instrument are

1. Low reading on the suction gauge
2. Possible warning flag on some instruments

Action required is to reerect the gyroscope. This is accomplished in flight by caging and uncaging the instrument when the aircraft is straight and level and at a constant speed to achieve an approximate reerection. However, if the suction reading is still low, the artificial horizon will still be unstable, and therefore, secondary flight instruments, e.g., turn coordinator, vertical speed indicator (VSI), etc., should be monitored.

The indication for a failed electrically driven artificial horizon is a power failure warning flag on the instrument face. In this circumstance, a backup power supply may be selected automatically or manually, but if this is not available, then the instrument will be unreliable and should not be relied on for accurate attitude information.

Describe the electrically driven artificial horizon instrument.

Electrically driven artificial horizons use the same basic principles as the air-driven instruments, with a gyroscope tied to the earth's vertical and two gimbals. (*See Q: Describe the air-driven artificial horizon instrument, page 137.*)

There are some fundamental differences, however, as follows:

1. Most gyroscopes/rotors of electrically driven artificial horizons rotate clockwise when viewed from above.
2. Electrically driven artificial horizons use an electric squirrel-cage motor to drive the rotor at approximately 22,000 rpm, which is about twice the speed of an air-driven instrument. Therefore, the electric artificial horizon is more rigid.
3. Electric erection systems are very fast, and because of this, there is no need for the gyroscope to be pendulous, although some do remain to some extent pendulous. Electric erection system also can be cut out at will.
4. Acceleration and turn errors are minimized or completely eliminated because the instrument has little or no pendulosity and its normal erection system can be cut out at certain values of longitudinal or lateral (balanced turn) acceleration.
5. The electrically driven artificial horizon has a freedom of pitch of ±85 degrees and unlimited roll.

What is a servo-driven attitude directional indicator (ADI) (or remote artificial horizon)?

A servo-driven, attitude directional indicator, also called a *remote artificial horizon,* is used on modern aircraft to display attitude information (pitch and roll) that has been calculated by the aircraft's inertia navigation/reference system (INS/IRS) platforms. (*See Q: What is an INS/IRS? page 92.*)

The system has the advantage of being free of turn and acceleration errors. Also, as a dual system (i.e., INS 1 feeds the captain's instruments, and INS 2 feeds the copilot's instruments), the displayed information can be compared, and the separate standby instrument can be used to detect errors and deselected incorrect systems.

Describe the turn and slip (turn coordinator) indicator instrument.

The turn and slip indicator is in effect two instruments combined as a single unit. One measures turn, and the other measures slip or skid.

Turn is the movement about the aircraft's yaw axis (the aircraft's vertical) that results in a change of direction.

Slip is a lateral force into the turn.

Skid is a lateral force out of a turn.

The turn and slip instrument consists of the following:

1. A rate gyroscope.
2. The gyroscope rotates about a horizontal axis.
3. One gimbal, which is pivoted about the aircraft's fore and aft axis that measures the aircraft's yaw when the precessed force in this plane of freedom is sensed.
4. Two planes of freedom:
   a. The gyroscope's spin axis.
   b. Yaw axis of the gimbals.
5. The gyroscope's axis is aligned to the aircraft's lateral axis.

Describe the following turn and slip indications.

(*See Figure 6.1, page 141.*)

Note: If the ball is to the left, use the left rudder. If the ball is to the right, use the right rudder. Hence the old adage, "kick or pressure the ball back into the center."

What errors does the turn and slip indicator experience?

The turn indicator gyroscope suffers from

1. Gyroscope system failures
2. Looping error. (This is an inherent design error in the instrument, and as a result, with any yaw condition the gyroscope will tilt.)
3. Real wander of the gyroscope's spin axis. (*See Q: What is real wander of a gyroscope? page 136.*)

What is the turn coordinator?

The turn coordinator is an advanced development of the earlier turn indicator. It is similar to the turn indicator instrument, except that its single gimbal is raised at the front by 30 degrees so that the instrument is sensitive to both roll and yaw, and it begins to indicate a turn as soon as the roll-in begins.

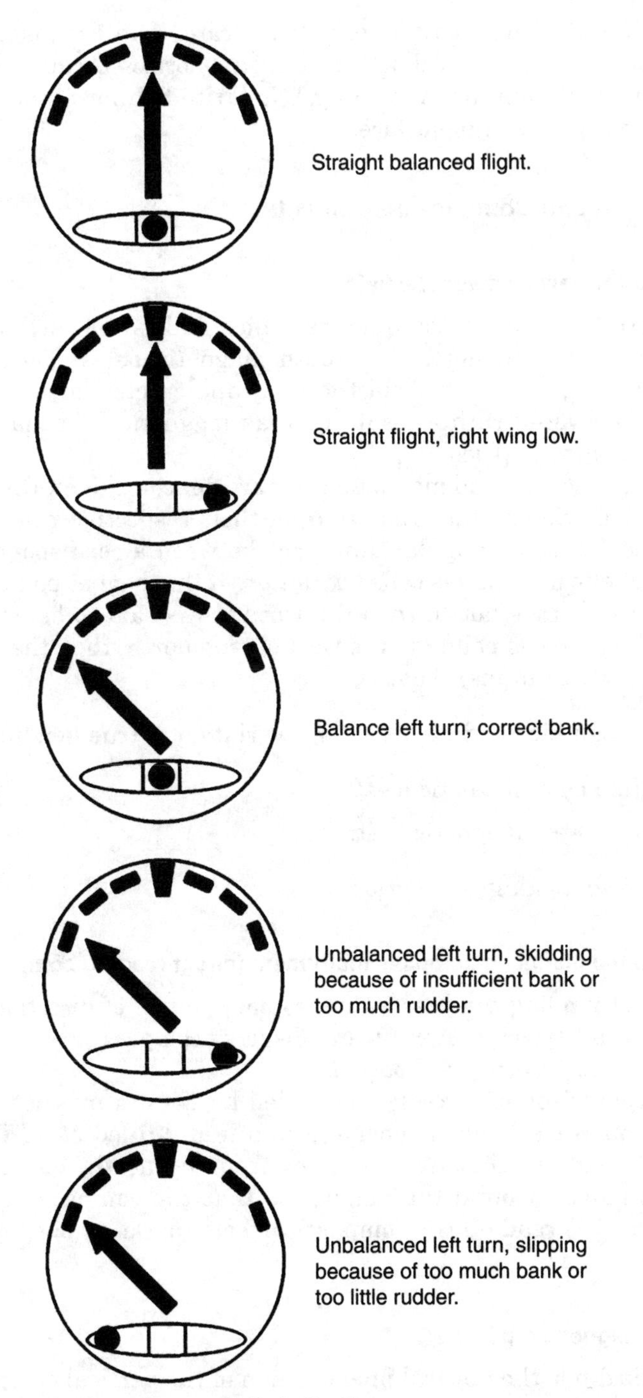

Figure 6.1 Turn and slip indications.

However, the turn coordinator only indicates rate 1 turns accurately and should not be confused with an artificial horizon because it displays no attitude information. A warning, "No Attitude Information," is often written on the instrument face.

Magnetism and Compass Instruments

Describe the earth's magnetic field.

The earth has a magnetic iron core, which makes the earth act like a giant magnet with north and south magnetic poles. The magnetic poles are slightly offset from the geographic poles, and the earth's surface is covered with a resulting weak magnetic field that radiates from its magnetic poles.

Because the true and magnetic poles are not coincident, the true and magnetic meridians that radiate from their respective poles are also not coincident. The angular difference between a corresponding true and magnetic meridian is called *variation.* If the magnet points slightly to the east of true north, then the variation is said to be east (plus), and if the compass points to the west of true north, then the variation is said to west (minus). That is,

Magnetic heading + easterly variation = true heading

Variation east magnetic least.

Variation west magnetic best.

(*See Q: What is magnetic variation? page 89.*)

Describe the magnetic compass instrument (direct reading compass).

The direct reading compass is the primary source of directional information in all types of aircraft and displays compass heading. (*See Q: What is compass direction? page 89.*)

It is comprised of a freely suspended horizontal magnet attached to a compass card that is enclosed in a liquid-filled case. The magnet will swing so that its axis points roughly north-south, and the aircraft moves around the magnet so that the compass heading of the aircraft is read off the compass card against a lubber line on the instrument case.

What is magnetic dip?

Magnetic dip is the natural phenomenon of the vertical component of the magnetic field over the earth's surface and its effect on the magnetic compass.

The earth's magnetic field has two components or forces: a horizontal force parallel to the earth's surface that is used to align the compass with magnetic north and therefore to determine direction and a vertical force that causes the needle to dip down.

At the magnetic equator, the horizontal force is dominant, and therefore, there is no dip and the compass is accurate. However, as you move closer to either of the magnetic poles, the vertical component increases, and this induces the magnetic bar in the compass to dip down to align itself vertically with the magnetic field.

Explain compass swinging.

Compass swinging is a procedure to check the accuracy of and to adjust an aircraft's magnetic compass.

A compass should be swung when any of the following occurs:

1. The compass is new.
2. Any equipment influenced by electrical or magnetic energy in the vicinity of the compass has been altered.
3. Having passed through a severe magnetic storm.
4. After a considerable change in latitude.
5. After any inspection of either
   a. The compass.
   b. Nearby equipment influenced by electrical or magnetic energy.
6. Whenever there is doubt about the accuracy of the compass.

Describe the errors of the magnetic compass.

The errors of the magnetic compass are

1. Acceleration/deceleration errors
2. Turning errors

Describe the remote indicating compass.

The remote indicating compass is a combination of the directional indicator (DI) and the magnetic compass instruments as a single instrument. It uses the rigidity of the gyroscope to avoid compass turning and acceleration errors and a magnetic north–sensing input to prevent DI gyroscope wander to maintain its correct orientation at all times without any external influence. (*See Qs: Describe the directional indicator and describe the magnetic compass, pages 136 and 142.*)

The remote indicating compass is made up of the following:

1. Detector unit
2. Gyroscope
3. Feedback system

Radio Instruments

Describe the relative bearing indicator (RBI) instrument and how it works.

The RBI is a simple automatic direction finder (ADF) instrument that is used to display nondirectional beacon (NDB) navigation information. The RBI is comprised of the following:

1. A fixed 360-degree compass card. The 000-degree position is at the 12 o'clock position and thus the aircraft heading.
2. An ADF needle that seeks out and shows the relative bearing of the NDB from the aircraft's heading.

The pilot, in conjunction with the directional indicator (DI), uses the RBI by adding the relative bearing (ADF needle) to the aircraft's magnetic heading (DI) to determine the QDM to the NDB.

A further development of the RBI instrument is the moving-card ADF, which can be orientated manually to the aircraft's heading. This subtle change means that the needle head now indicates the QDM to the NDB.

Describe the radio magnetic indicator (RMI) instrument and how it works.

The RMI can be used to display automatic direction finder (ADF) or VHF omni range (VOR) navigation information and is regarded as an advanced development of the RBI.

The RMI is comprised of the following:

1. A remote indicating 360-degree compass card that is continuously and automatically aligned with magnetic north. Therefore, the RMI displays the aircraft's magnetic heading at the top of the dial.
2. Either a single or dual needle that seeks out the direction of the station to which the pilot is tuned and is superimposed onto a compass card that is orientated to the aircraft's magnetic heading. This means that the needle's head indicates a QDM and the needle's tail indicates a QDR.
3. Selection button, sometimes known as rabbit ears, enables the pilot to change between ADF and VOR needle indications.

   ADF selection. Selected as an ADF needle, it seeks out the direction of a nondirectional beacon (NDB) station as a QDM. The relative bearing of the beacon from the aircraft is the sum of the QDM minus the aircraft heading.

   VOR selection. Selected as a VOR needle, the head indicates the QDM and the tail indicates the QDR, which itself is an indication of the VOR radial the aircraft is on.

Describe the omni bearing indicator (OBI) instrument.

The OBI indicator is a navigation instrument sometimes referred to as a *first-generation VHF omni range (VOR) indicator.* It is used by the pilot to select the required VOR radial and to display tracking guidance relative to the selected radial. If the aircraft is on the selected radial, the VOR needle or course deviation indicator (CDI) will be centered, and if the aircraft is not on the selected course/track, then the CDI will not be centered.

The OBI is a track-up display; i.e., the selected track accommodates the top of the dial position regardless of the aircraft's heading.

Note: Later versions of the OBI instrument also have the capability of displaying instrument landing system (ILS) information.

Note: *Course* is another term for *track.*

Describe the purpose-built instrument landing system (ILS) indicator.

A purpose-built ILS display instrument is a further development of the omni bearing indicator (OBI) instrument. It still acts as a VHF omni range (VOR) navigation display when a VOR frequency is selected, but it also can act as an ILS display instrument when an ILS frequency is selected to guide a landing aircraft along both a localizer track and a glide slope descent path.

The ILS indicator is a track-up display, like the OBI instrument. That is, the selected track (localizer) accommodates the position at the top of the dial regardless of the aircraft's heading. The ILS indicator is similar to the OBI instrument [see Q: *Describe the omni bearing indicator (OBI) instrument, page 145*] with two particular differences:

1. *Localizer deviation bar and dot scale.* The track deviation bar (vertical needle) and the horizontal dot scale represent the aircraft's horizontal angular deviation from the ILS localizer beam. One dot represents an ILS localizer deviation of ½ degree. A full five-dot scale represents a total ILS localizer deviation of 2½ degrees or more.
2. *Glide slope deviation bar and dot scale.* The ILS instrument also displays the aircraft's angular deviation from the ILS glide slope beam using a horizontal needle/bar superimposed on a vertical dot scale in the middle of the display. The glide slope needle/bar shows the position of the actual glide slope, and the center of the dot scale represents the aircraft's relative position to the glide slope. Thus the pilot's aim is to maintain the glide slope bar/needle on the center dot of the glide slope scale. The measurement of the glide slope dot scale is as follows: One dot represents a glideslope deviation of 0.15 degree.

A full dot scale deviation represents a glide slope deviation of 0.7 degree or more. The glide slope display is activated automatically once the ILS has been tuned and selected.

Both the localizer and the glide slope displays should be used as command instruments. That is, fly toward the deflected needles to center both the localizer and glide slope tracks as a center cross display.

Describe the horizontal situation indicator (HSI) instrument.

The HSI is a sophisticated primary navigation instrument. It is comprised of a remote indicating compass that displays the aircraft's directional magnetic heading and an instrument landing system/VHF omni range (ILS/VOR) display that gives an easy-to-understand display of the aircraft's situation in relation to the selected VOR radial or ILS localizer and glide slope and the aircraft's magnetic heading. Because the basis of the HSI is the remote indicating compass, the instrument is a head-up display; i.e., the aircraft's magnetic heading accommodates the position at the top of the dial regardless of the aircraft's selected course.

The mode, either ILS or VOR, that the HSI works to is determined automatically by the VHF/NAV signal selected and received by the aircraft. That is, selecting a VOR frequency, the HSI displays VOR indications, etc.

The HSI instrument is found in most modern aircraft and consists of the following:

1. Remote indicating compass
2. Combined course/track and deviation bar
3. Localizer dot scale
4. Glide slope dot scale

The HSI always acts as a command instrument because the head-up display of the RIC ensures the correct orientation at all times whenever the HSI is tuned to a VHF signal, either VOR or ILS. By flying toward the deflected deviation bars, the aircraft can center the course/track and glide slope bars and thereby regain track.

For direction-measuring equipment (DME) questions, see subchapter, "Ground Navigation Systems," pages 107–109.

Describe a radio altimeter and how it works.

Radio altimeters provide an accurate height measurement from 2500 ft down to 50 ft above ground level (AGL) for pulse radar beams or 0 ft for continuous-wave radar beams. They are usually fitted alongside barometric altimeters in most commercial aircraft.

The basic principle of the radio altimeter is that a wide conical beam is directed vertically down toward the ground, and the time taken for the reflected signal to return corresponds to its height.

At what height does a radio altimeter normally become active?

A radio altimeter normally becomes active at 2500 ft above the ground for both separate dial and electronic flight instrument system (EFIS) radio altimeter instruments.

Note: Some types of radio altimeters may become active at a different height.

Advanced Flight Instruments

What does EFIS stand for?

Electronic flight instrument system.

What is an electronic flight instrument system (EFIS)?

EFIS is a fully integrated computer-based digital navigation system that uses color cathode-ray tube (CRT) types of electronic attitude directional indicator (EADI) and horizontal situation indicator (EHSI).

What components make up a typical EFIS?

An electronic flight instrument system has the following five main components:

1. Cathode-ray tubes (CRTs)
2. EFIS control panel
3. Symbol generators
4. EADI (electronic attitude directional indicator)
5. EHSI (Electronic horizontal situation indicator)

What is the advantage of an EFIS flight deck?

The electronic flight instrument system (EFIS) display has two distinct advantages over older, mechanically driven instruments. First, it displays the same information in a clearer and more versatile manner, and second, it can bring together additional data from several different sources to present the pilot with the best possible attitude and navigation information for a particular stage of flight on a dual or single display panel.

What is typically displayed on the EADI?

The electronic attitude directional indicator (EADI) display includes the following:

1. Basic attitude information (pitch and roll) and a turn and slip indicator (yaw) received from an inertia reference system (IRS)
2. Additional attitude information
   a. Flight director command bars
   b. Pitch limit symbols, also known as *eyebrows*
   c. Rising runway
3. Speed indicator:
   a. Speed tape (side of EADI)
   b. Fast/slow speed indicator (speed trend)
   c. Mach number and ground speed digital display
4. Navigation information:
   a. L Nav or localizer deviation indicator (bottom of EADI)
   b. V Nav or glide slope deviation indicator (right-hand side of EADI)
5. Altitude, radio altimeter height, and decision height display
6. Autopilot, armed and engaged modes:
   a. Autothrust
   b. Pitch mode
   c. Roll mode
   d. Autopilot status

At what height would you expect the rising runway symbol on an electronic attitude directional indicator (EADI) to become active?

The rising runway normally becomes active at 200 ft radio altimeter, but this can vary because it is a type-specific design feature.

How is 0 ft represented by the rising runway on the electronic attitude directional indicator (EADI)?

Zero feet is represented by the rising runway symbol reaching the base of the aircraft symbol.

For flight director questions, see "Flight Management and Auto Flight System," page 160.

What are the electronic horizontal situation indicator (EHSI) instrument modes?

The EHSI typically has seven display modes, which are as follows:

1. Full VOR/ILS
2. Full NAV

3. Expanded (arc) VOR/ILS
4. Expanded (arc) NAV
5. Map mode
6. CTR map mode
7. Plan

Which electronic horizontal situation indicator (EHSI) modes can display the weather radar?

Weather radar typically can be overlaid on the following modes:

1. Expanded VOR/ILS
2. Expanded NAV
3. Map mode
4. Center map mode

What are the general electronic flight instrument system (EFIS) color coding?

There is no standard color coding used by all the different EFIS manufacturers. However, in general, the following color scheme is the most common:

Green	Active or selected mode, changing conditions
White	Present situation and scales
Magenta	Command information and weather radar turbulence
Cyan	Nonactive background information
Red	Warning
Yellow	Caution
Black	Off

What does HUD stand for?

Head-up display. This normally consists of electronic attitude directional indicator (EADI) information, i.e., speed, attitude, and flight director bars, etc.

What does HUGS stand for?

Head-up guidance display. This normally consists of electronic attitude directional indicator (EADI) information and additional navigation guidance raw material, i.e., localizer and glide slope.

Note: HUGS allows Cat III instrument landing system (ILS) landings only on runways certified for Cat I landings.

Radio Communication Systems

What do you know about very high frequency (VHF) communications?

VHF radio transmissions are used for short-range communications. VHF radio uses line-of-sight propagation paths and allows reception and transmission at any point within its area of coverage, namely, from the ground station to its maximum range.

Civil agencies uses ultrahigh frequency (UHF) in the 118- to 137-MHz band at 12.5-kHz intervals. This range usually gives good reception and only slight interference from static tied to atmospheric attenuation.

What is the range of a VHF signal, e.g., at 35,000 ft?

A VHF signal uses line-of-sight propagation paths.

$$\text{Line-of-sight formula} = 1.25\sqrt{H_1 \text{ (in feet AMSL)}} + 1.25\sqrt{H_2 \text{ (in feet AMSL)}}$$

where H_1 is the height of the transmitter and H_2 is the height of the receiver. Thus,

$$1.25\sqrt{0} + 1.25\sqrt{35{,}000} = 234 \text{ nautical miles}$$

What factors affect the range of VHF communications?

The following factors affect the range of a VHF communication:

1. Transmitter power
2. Frequency
3. Height of transmitter and receiver
4. Obstructions
5. Fading

What do you know about high-frequency (HF) communications?

HF radio transmissions are used for long-distance communications between two specific points only, unlike VHF. It uses predictable sky wave propagation paths that are refracted off the earth's ionosphere over great distances.

The HF band used in aviation ranges from 2 to 22 MHz. Different frequencies have different range capabilities; i.e., the higher the frequency, the greater is its range. Therefore, a frequency is chosen by the pilot to meet the range between the transmitting and reception points, and the ground station will monitor a range of frequencies because personnel are unsure of the exact distance to the aircraft. This is also the reason why you can be received by a station 2000 nautical miles away but not by a station only 500 nautical miles away.

How are HF communications affected at night (winter)?

At night, half the HF of the daytime frequency produces the same range (skip distance). This is so because of variation in the ionosphere. During the day, especially in the summer, the sun generates ion particles that make up the ionosphere's D layer at a height of approximately 75 km. This layer is of sufficient density to refract HF sky waves. However, at night or during winter days when the exposure to the sun is less, the D layer disappears, and therefore, the HF sky waves are refracted by the ionosphere's E layer at a height of approximately 125 km. This increases the range (skip distance) of an HF transmission because of the greater vertical distance to the higher E layer before it is refracted.

Therefore, because of the higher ionosphere at night, you need a lower frequency to reach the same receiver distance; typically, half the daytime frequency is needed because the signal is refracted more.

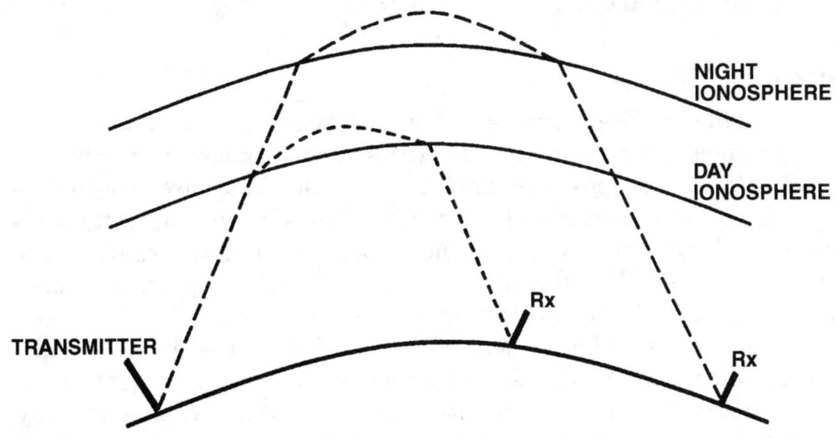

Figure 6.2 Day/night variation of ionosphere sky-wave refraction.

What factors affect the range of HF communications?

The following factors affect the range of an HF communication:

1. Transmitter power
2. Frequency
3. Time of day
4. Season
5. Location
6. Disturbance of the ionosphere

Avoidance Systems

What is TCAS?

Traffic (Alert) Collision Avoidance System (TCAS) provides traffic information and maneuver advice between aircraft if their flight paths are conflicting with each other.

TCAS uses the aircraft's secondary surveillance radar (SSR) transponders and is completely independent of any ground-based radar units. TCAS is rapidly becoming a mandatory requirement around the world and is already established in U.S. airspace.

TCAS I is an early system that provides traffic information only.

TCAS II is a later system that provides additional maneuver advice but in the main is restricted to vertical separation.

TCAS IV is a new system under development (1998) that will give resolution advisories (RAs) in the horizontal and vertical planes. However, further development of TCAS IV is likely to be canceled in preference to ADS-B.

How does TCAS work?

An aircraft's traffic collision avoidance system (TCAS) will interrogate the secondary surveillance radar (SSR) transponders of nearby aircraft to plot their positions and relative velocities. Direction-finding aerials obtain the relative bearings of other aircraft, and distance is calculated by using the time delay between the transmitted and received signals. With this information, the TCAS computer can determine the track and closing speeds of other aircraft fitted with transponders, and where it determines a collision is possible, it provides visual and aural warnings as well as command actions on how to avoid the collision. This is all done with vertical avoidance commands only. As yet, no turn commands are given.

The warnings and advice (advised actions) from the system have different levels:

Initially. A traffic advisory (TA) warning is generated for other traffic that may become a threat. No maneuver is advised or should be taken.

Collision threat. A resolution advisory (RA) warning is generated when an aircraft is considered to be on a collision course. Advice on a maneuver in the pitching plane, i.e., rate of climb or descent, to avoid the collision is generated that can be increased or decreased as the threat increases or reduces until a clear of conflict notice is given.

Therefore, you only respond to an RA, which should be done promptly and smoothly and should take precedence over air traffic control (ATC) clearance to avoid immediate danger.

RA use should be restricted in the following circumstances:

1. In dense traffic area (limited to TA use)
2. Descent recommendations inhibited below 1000 ft
3. All RAs inhibited below 500 ft (*Note:* All TAs also restricted below 400 ft)

TCAS can cope with mode A, C, and S transponders. However, when both aircraft are equipped with TCAS II and mode S, the advice on how to avoid a collision will be coordinated by the mode S data link between the two aircraft.

What is ACAS?

ACAS is a European airborne collision avoidance system (ACAS II).

What is ADS-B?

Automated Dependent Surveillance–Broadcast (ADS-B) system is a traffic collision avoidance system similar to TCAS. ADS-B incorporates cockpit display of traffic information and is overall a more capable system than TCAS II.

What is GPWS, and how does it work?

Ground proximity warning system.

The GPWS is essentially a central computer system that receives various data inputs on configuration, height/altitude, and instrument landing system (ILS) glide slope deviation. (*See Q: What are the inputs to a GPWS? page 154.*) It then calculates these inputs to detect if any of the following dangerous and/or potentially dangerous circumstances exist:

1. Excessive barometric rate of descent
2. Excessive terrain closure rate
3. Height loss after takeoff
4. Flaps or gear not selected for landing
5. Too low on the ILS glide slope
6. Descending below approach minima

These circumstances form the six main working modes of the GPWS.

Each mode has a precomputerized active range with distinct boundaries. Most modes have two boundaries:

1. Initial alert boundary, whereby a potential danger to the aircraft's safety exists.
2. Warning boundary, whereby a present danger to the aircraft's safety exists. A warning usually follows an alert if the condition persists. However, some modes (5 and 6) do not have a warning boundary, only an alert. (*See Q: What are the initial actions required for a GPWS alert and warning? page 156.*)

If the central computer determines that a boundary of any of the various modes has been exceeded, it will activate aural and visual alerts to warn the flight crew using the following cockpit equipment:

1. Speakers.
2. A pair of red warning lights with "Pull Up."
3. A glide slope deviation inhibited amber advisory light.

When more than one alert or warning is triggered at the same time, the GPWS will give only the highest-priority alert or warning that has been triggered. (*See Q: What is the GPWS modes order of priority? page 155.*)

The GPWS is now mandatory equipment for public transport aircraft over 5700 kg.

What is EGPWS?

Enhanced ground proximity warning system.

Initially, EGPWS was only suitable for use on aircraft with digital avionics, although research is ongoing to develop an analogue aircraft version. EGPWS provides a greater level of detection than a standard GPWS. For example, terrain mapping is a new feature on EGPWS. It can be shown on navigation displays by using the weather system. Probable windshear aural and visual warnings also can be generated to warn of an impending possibility of encountering windshear ahead.

What are the inputs to a GPWS?

The inputs to a ground proximity warning system (GPWS) are as follows:

1. Barometric altitude for rate of descent (ROD) calculations
2. Radio altimeter
3. Flap position
4. Gear position

5. Instrument landing system (ILS) glide slope
6. Approach minima
7. Throttle position

What are the various GPWS modes?

There are six main ground proximity warning system (GPWS) modes, of which some have further subdivisions:

1. Mode 1. Excessive barometric rate of descent
2. Mode 2. Terrain closure
3. Mode 3. Sinking flight path after takeoff or go-around
4. Mode 4. Gear and flap not selected
5. Mode 5. Instrument landing system (ILS) glide slope deviation
6. Mode 6. Approach minimas

Windshear warning is sometimes classed as a GPWS mode because it uses the GPWS flight crew visual and aural warning system. However, strictly speaking, it is not a ground proximity warning but a separate system. Because it is often perceived as a GPWS mode, we shall classify it as both an additional GPWS mode (mode 7) and a separate warning system. (*See Q: Describe windshear warning, page 156.*)

7. Mode 7. Windshear warning

When triggered, most of the modes have two boundaries:

1. *Alert.* This is followed by
2. *Warning.* If the condition persists.

(*See Q: What are the initial actions required for a GPWS alert and warning? page 156.*)

What are the GPWS modes order of priority (not including windshear)?

Highest priority	Whoop, whoop, pull up (Modes 1, 2, 3, 4)
	Terrain, terrain (Mode 2)
	Too low gear (Mode 4a)
	Too low flap (Mode 4b)
	Minima, minima (Mode 6)
	Sink rate, sink rate (Mode 1)
	Don't sink, don't sink (Mode 3)
Lowest priority:	Glide slope, glide slope (Mode 5)

What are the initial actions required for a GPWS alert and warning?

For a ground proximity warning system (GPWS) alert, the initial action required is a corrective response; i.e., "Too low flaps" requires the corrective action of lowering the flaps accordingly. For a GPWS warning, the immediate action required is a full windshear go-around, i.e., maximum go-around thrust and climb attitude.

When can you override a GPWS warning?

The captain can decide to override a ground proximity warning system (GPWS) warning when the aircraft is

1. 1000 ft vertically clear of clouds
2. 1 km horizontally clear of clouds
3. In 8½ km of clear visibility
4. Flying in the daytime
5. Obviously not in danger

Describe windshear warnings.

Modern aircraft are provided with windshear warnings by using air data computer–detected changes in airspeed to calculate the presence of a windshear phenomenon, which it feeds to the ground proximity warning system (GPWS), which generates a "Windshear, windshear" aural and visual display on the main attitude directional indicator (ADI) flight instrument. Windshear warning systems are active from the ground level to a height of 1500 ft.

A windshear warning requires an immediate go-around at full thrust and maximum flight director pitch-up attitude to avoid ground contact. If, however, the aircraft still does not climb, then the aircraft should be pitched up to its maximum pitch limit symbols or eyebrows. (*See Q: What is windshear? page 254.*)

What is the pilot's order of priority given a windshear, GPWS, and TCAS warning at the same time?

First: Windshear

Second: Ground proximity warning system (GPWS)

Third: Traffic collision avoidance system (TCAS)

For weather radar questions, see Chapter 4, "Navigation," subchapter, "Radar," pages 112–113.

Flight Management and Auto Flight Systems

What is the purpose of a flight management system (FMS)?

The purpose of an FMS is to manage the aircraft's performance and route navigation to achieve the optimal result. (*See Q: Describe a typical flight management system, page 157.*)

These managed data then can be used for either (1) advisory crew information only or (2) to direct the autothrottle and autopilot to steer the aircraft. (*See Q: Describe an autopilot system? page 158.*)

What are the flight management system's (FMS's) three sources of input data?

The three sources of flight management system data are

1. Stored databases
2. Pilot inputs via the control display unit (CDU)
3. Other aircraft systems, which are fed automatically into the FMS

These input sources are fed into the flight management computer (FMC) and are duplicated for both performance and navigation management functions.

Describe a typical flight management system.

This is really a type specific question, and the reader should refer to his or her own aircraft type; however, the following answer, which is based on the B737-300 FMS, is given as an example.

The flight management system (FMS) combines data from various aircraft systems, which are fed into a central computer (FMC) to manage the route navigation and the aircraft's performance. (*See Q: What are the flight management system's (FMS's) three sources of input data? page 157.*)

The main section of the FMS is the flight management computer system (FMCS), which is a dual system using two flight management computers (FMCs) and two control display units (CDUs).

The CDUs are the single point of control used by the crew to access the FMC. Via the CDUs, the crew can initiate and control a given flight plan's navigation and performance management and monitor its execution. The CDU also provides a convenient access point to initiate the inertia reference system (IRS) units. Either CDU can input to both FMCs simultaneously, and each FMC will then compute and execute the command.

The CDU consists of a monochrome cathode-ray tube (CRT) screen and a keyboard for data entry. The central part of the screen is used to display selected data. Below the data block is a scratchpad area where FMC messages can be shown and where data are held before being entered into the system.

Mode select keys are used to control the type of data available on the CDU, e.g., climb, route, and fix.

There are two main operational functions of the FMC/CDU:

1. Route navigation management
2. Aircraft performance management

What are the L NAV and V NAV flight management computer (FMC) functions?

The L NAV and V NAV functions are predetermined or custom-made route profiles. They are based on the FMC's managed performance and navigation data and are stored in the FMC databases so that they can be selected via the control display unit (CDU) to be flown either as an autopilot and autothrottle modes of operation or as a guide for manual flying.

The L NAV (lateral navigation) function guides the aircraft's lateral movement and is available from takeoff to localizer capture, and the V NAV (vertical navigation) function guides the vertical path of the aircraft, including climb and descent profiles.

A stored company route will contain a complete lateral and vertical flight profile, which can be amended by the crew if required.

Describe an autopilot system.

An autopilot system is an integrated flight control system that enables an aircraft to fly a prescribed route and to land at a designated airport without the aid of a human pilot.

What is the purpose of an autopilot system?

The purpose of an autopilot system is to relieve the pilot of the physical and mental fatigue of flying the aircraft, especially during long flights. This will result in the pilot being more alert during the critical phase of landing the aircraft, thus improving safety.

What work functions does an autopilot achieve?

When engaged, the autopilot is responsible for the following:

1. Stabilizing the aircraft
2. Maneuvering the aircraft

For practical stabilization reasons, the autopilot is broken down into three basic control systems:

Pitch. To control the elevators.

Roll. To control the ailerons.

Yaw. To control the rudder.

The autopilot's maneuver capabilities are classified by the number of axes the autopilot controls:

Single axis. A single-axis autopilot commands the ailerons that control the aircraft about the roll axis only.

Two axes. A two-axis autopilot commands the ailerons and elevator that control the aircraft about the roll and pitch axes.

Three axes. A three-axis autopilot commands the ailerons, elevator, and rudder that control the aircraft about the roll, pitch, and yaw/vertical axes.

How does the response/reaction of an autopilot system compare with that of a pilot?

The response of an autopilot is substantially faster than that of a human pilot.

A human pilot takes approximately ⅕ second in detecting a change in the aircraft's attitude and then a further delay in deciding which control to apply to oppose the disturbance. An autopilot, however, will detect a disturbance and put on the required control to correct the disturbance in approximately 50 ms.

Explain the various autopilot modes of operation.

This is really a type-specific question, and the reader should refer to his or her own aircraft type; however, the following answer, which is based on the B737-300 autopilot system, is given as a example.

The autopilot modes of operation are normally the following:

1. Heading
2. Lateral navigation (L NAV)

   Note: This is known as managed navigation for Airbus aircraft.

3. VOR/LOC (VOR/localizer mode)
4. Altitude hold

5. Vertical speed
6. Level change

Note: This is known as *open climb or descent* for Airbus aircraft.

7. Vertical navigation (V NAV)

Note: This is known as *managed* vertical navigation for Airbus aircraft.

8. ILS/approach

Note: *For speed and thrust modes, see Q: Explain the autothrottle operating control modes, page 164.*

Describe the flight director system.

The flight director system (FDS) provides guidance information on the required aircraft maneuver in pitch and roll to gain, regain, or maintain the programmed flight path.

The flight director display is found on the primary flight instrument (ADI) as either a chevron indicator or a pair of vertical and horizontal bars. Both the chevron and bar flight director systems are command displays; i.e., by flying toward the deflected bars, they will center to gain, regain, or maintain the programmed flight path.

The FDS uses the autopilot (AP) sensors and the aircraft's flight management computer (FMC) performance database to compute the aircraft's pitch and roll response to gain, regain, or maintain the programmed flight path, and the flight control computer (FCC) positions the FDS bars/chevron accordingly on the respective attitude directional indicators (ADIs).

What errors does a flight director system (FDS) have?

The errors of an FDS are primarily operating errors.

First, an error commonly occurs with interpretation of the FDS indication by the pilot, in that a pilot interprets the flight director bars as the flight path (e.g., center and crossed flight bars as meaning that the aircraft is on track). However, such a flight director indication also can and often does mean that the aircraft is on the correct intercept course. Therefore, such errors of interpretation are associated with FDSs, although they are strictly a pilot error.

Second, the FDS is part of the autopilot system, and because of this, the flight director bars are geared to the autopilot's quick response/reaction rate, which is far quicker than the ability of a human pilot. This therefore gives rise to problems when a pilot, especially on an approach, is flying the aircraft manually. That is, a small localizer deviation will command a large correcting bank by the flight director vertical bar to regain the localizer, a maneuver that the autopilot's quicker response rate can cope with but the human pilot is too slow in his or her reaction when removing the bank and therefore cannot keep up with the flight director bar. This results in the aircraft going through the localizer, and from there on, the aircraft will oscillate on either side of the centerline. Therefore, this can be seen as a true flight director system error when used for manual flying maneuvers; however, awareness of the problem should allow the pilot to overcome this error by using raw data in his or her maneuver decision making.

Describe an autoland system.

An autoland system is a function of the autopilot flight director system (FDS) and autothrottle system that is engaged using the *approach* mode of operation. It is designed to carry out automatic landings under all visibility conditions by providing better guidance and control than that provided by a pilot from the interception of the localizer until the touchdown.

The autoland system controls the aircraft about all three axes simultaneously. This includes the use of yaw to control drift and use of the autopilot's flight control computers, including sensors and instrument landing system (ILS) radio couplings, to accurately track the localizer and glide path approach down to minimas that can include a full Cat III autoland. (*See Q: Describe an autopilot system, page 158.*)

The autothrottle is also used by the autoland system to control and maintain the correct airspeed through engine power changes during the approach, to retard thrust during the flare, and finally to automatically disconnect at touchdown (type-specific).

What is a multiplex autoland system?

Multiplex is a term used for an autopilot landing system that is comprised of two or more independent autopilots/channels that are used collectively to provide a redundant system design.

A multiplex autoland system can use either

1. A dual digital control computer channel known as a *duplex system*
2. A triple digital control computer channel known as a *triplex system*
3. A quadruple digital control computer channel known as a *quadruplex system*

The independent auto pilot/channel systems are interconnected, and together they provide continuous autoland control. In the event of a single autopilot/channel failure or a different reading, that system is outvoted by the others and is disengaged automatically.

This multiplex redundancy system allows the autopilot to operate in either a fail passive or operational condition depending on the number of channels engaged, thus allowing the system to perform what are termed *land 2* or *land 3* status autolands. (*See Q: What is a fail passive/operational autopilot/landing system? page 162.*)

What is a fail passive autopilot/landing system?

A fail passive automatic pilot landing system, also known as a *land 2* system, is one that employs two digital control computer channels (duplex system). In the event of a single control channel failure, there is no significant out-of-trim condition or deviation of the flight path or attitude. However, the landing is not completed automatically due to a minimum dual channel system being required for a fully automated landing, thus requiring the pilot to assume control of the aircraft to complete the landing.

What is a fail operational autopilot/landing system?

A fail operational automatic pilot landing system, also known as a *land 3* system, is one that employs three digital control computer channels (triplex system). These provide redundant operational capability, whereby in the event of a single control channel failure below the alert height, the approach, flare, and landing can be completed by the remaining automatic systems due to the minimum required dual channel system still being available. This allows the automatic landing system to work in a fail operational manner.

Describe the autothrottle control system.

This is really a type-specific question, and the reader should refer to his or her own aircraft type; however, the following answer, which is based on the B737-300 autothrottle system, is given as an example.

Autothrottle (AT) systems are designed to control and maintain thrust and/or airspeed, especially during an automatic approach and landing, by changing engine thrust. They also may provide a constant closure rate of the throttle levers during the autoflare phase on some designs.

The autothrottle is part of the autopilot (AP) flight director system (FDS) and is available with or without the AP or FDS being engaged as long as a thrust or speed operating mode has been engaged.

An AT computer receives inputs from various sources, especially (1) the flight management computer (FMC), which calculates EPR/N_1 limits and targets and (2) APFDS pitch and speed targets and limits, which it uses to calculate the engaged operating mode's engine thrust targets and associated throttle lever positions (which are moved and held by servo driven motors and a clutch system) to attain the desired engine thrust level. (*See Q: Describe the autothrottle operating control modes, page 164.*)

How do you engage an autothrottle system?

This is really a type-specific question, and the reader should refer to his or her own aircraft type; however, the following answer, which is based on the B737-300 autothrottle system, is given as an example.

The autothrottle is engaged by a master switch on the mode control panel (MCP) being selected to "arm" and then an operating mode being selected either automatically by the autopilot or manually through switches on the MCP.

How can you disconnect an autothrottle system, and what indications are there?

This is really a type-specific question, and the reader should refer to his or her own aircraft type; however, the following answer, which is based on the B737-300 autothrottle system, is given as an example.

The following can disconnect the autothrottle (A/T):

1. The pilot can override the AT by applying normal pressure to the throttle levers. This applies an opposing force to the servo drive, which will automatically disengage the clutch.
2. The AT disengage switch located on the end of each throttle lever can be pressed.
3. The AT engagement switch located on the mode control panel (MCP) can be set to off.
4. The AT automatically disengages after touchdown and for various type-specific abnormal situations.

What are TOGA switches?

This is really a type-specific question, and the reader should refer to his or her own aircraft type; however, the following answer, which is based on the B737-300 autothrottle system, is given as an example.

The takeoff go-around (TOGA) switches (or gate on an Airbus aircraft) provide a means of engaging the autothrottle and the flight director in the takeoff or go-around mode. The TOGA switches are located on the aft edge of each throttle lever.

Explain the autothrottle operating control modes.

This is really a type-specific question, and the reader should refer to his or her own aircraft type; however, the following answer, which is based on the B737-300 autothrottle system, is given as an example.

The normal autothrottle's modes of operation are

1. Takeoff thrust
2. Go-around thrust
3. Maximum continuous thrust
4. Airspeed
5. Airspeed/Mach hold

How does the autopilot flight director system (APFDS) and the autothrottle (AT) combine to control attitude and speed during (a) the takeoff, (b) the climb, (c) the cruise, (d) the descent, and (e) the approach?

This is really a type-specific question, and the reader should refer to his or her own aircraft type; however, the following answer, which is based on the B737-300 autothrottle system, is given as an example.

The autothrottle and APFDS work together to maintain the aircraft's airspeed and vertical path in different variations for the phase of flight and/or the mode of operation selected.

1. *Takeoff.* The AT sets takeoff thrust, and the APFDS adjusts pitch to maintain the airspeed, V_2 + 10–20 knots.
2. *Climb.* Either
   a. The APFDS controls airspeed with pitch commands, and the AT controls engine thrust to a specific N_1 climb thrust value (e.g., level change or V NAV climb).
   b. The APFDS controls the vertical path attitude, and the AT maintains airspeed through the engine thrust control (e.g., vertical speed).
3. *Cruise.* The AT controls engine thrust to maintain airspeed, and the APFDS controls attitude to maintain altitude.
4. *Descent.* Either
   a. The throttles retard to idle to gain the attitude that gives the maximum descent rate, and the APFDS controls the pitch to maintain airspeed (e.g., level change/V NAV descent).
   b. The AT controls airspeed, and the APFDS controls the vertical flight path (e.g., vertical speed/V NAV path descent).
5. *Approach.* The AT controls airspeed, and the APFDS controls the attitude to maintain a vertical path profile (e.g., autoland *approach* mode).

Aircraft Systems

Technical questions asked in an interview about aircraft systems tend to be type-specific. Thus knowledge of your particular aircraft is required, and as such, only general questions are covered in this section.

What aircraft systems use hydraulic power?

The common aircraft systems that use a hydraulic force are

1. Landing gear
2. Brakes/antiskid system
3. Steering
4. Flying controls:
   a. Ailerons
   b. Elevator
   c. Rudder and yaw damper
   d. Trailing-edge flaps
   e. Leading-edge flaps and slats
   f. Spoilers:
      (1) Flight
      (2) Speedbrakes
5. Stairs
6. Doors

What are the advantages of a hydraulic system over a pneumatic system?

The advantages of a hydraulic system over a pneumatic system are as follows: Hydraulic fluid is incompressible, and this makes the system respond instantly and more efficient. In addition, hydraulic fluid that leaks from the system is easier to detect visually than air from a pneumatic system.

What are the advantages of a pneumatic system over a hydraulic system?

The advantages of a pneumatic system over a hydraulic system are as follows:

1. Air weighs less than hydraulic fluid, which is beneficial to the aircraft's overall weight limitations.
2. You do not have any problems of availability and cost with air, which you could have with hydraulic fluids.

What is a typical landing gear layout?

Most aircraft use the tricycle layout, where the two main undercarriage units are positioned just aft of the center of gravity and support up to 90 percent of the aircraft's weight and the entire initial landing shocks. The nose-wheel unit keeps the aircraft level and in most cases provides a means of steering.

What is the purpose of retractable landing gear?

The main purpose of retractable landing gear is to improve aircraft performance by reducing the drag created by extended gear in flight.

How is the landing gear extended and retracted?

The gear is operated by the crew member via a gear lever in the cockpit, and a system of green and red lights for each wheel indicate whether the associated wheel is down and locked (green) or up and locked (red).

Gear retraction normally is accomplished by a hydraulic system, although pneumatic or electrical systems also can be used. Often a hydraulic system is the primary source of gear retraction power, and an electrical system or an emergency accumulator hydraulic power source of gear retraction is used as a backup system. In all cases, the retracted gear is secured in its stowage position by mechanical locks.

The gear usually is extended using the power source that retracts the gear, i.e., hydraulics, although sometimes the gear is extended simply by releasing the mechanical locks and allowing gravity to extend the gear. Once the gear is down, mechanical locks hold it in position. However, in the main, gravity extension is used only as a backup procedure.

In addition, a means is normally provided to protect from retracting the gear on the ground, via a ground/air sensor that locks the gear lever in the down position when the aircraft is on the ground. However, an override switch (normally on the underside of the gear lever) is provided for emergency use when "belly flopping" is a better option in an unusual situation.

What is nose wheel shimmy?

Nose wheel shimmy is unstable swiveling oscillation of the nose wheel due to the flexibility of tire sidewalls, especially at high speed. Excessive shimmy can vibrate dangerously throughout the entire aircraft, causing wear in the wheel bearings, low tire pressures, and wear on the undercarriage linkage and mountings.

What are the requirements of a nose wheel?

The nose wheel requirements are as follows:

1. To carry the aircraft's direct compression load weight
2. To provide a towing attachment

3. To withstand shear loads
4. Castoring (The nose wheel must be able to castor freely when subjected to compression and shear loading; i.e., right brake to turn right.)
5. Self-centering (Whenever the weight is remove from the wheel, it must be centered to ensure correct positioning of the wheel before retraction.)
6. Steering
7. Antishimmy

What is tire creep?

Creep is the tendency of the tire to rotate slowly (creep) around the wheel hub as a result of a millisecond landing friction on the tire before wheel spin occurs (usually because of a low tire pressure). This creep, if excessive over a period of time, will cause the tire to tear out the inflation valve and cause the tire to burst on touchdown.

How is creep detected?

When the tire is installed a mark, usually a red dot, is placed on the tire sidewall, and a second dot is placed adjacent on the wheel hub. From then on, any displacement of the red dots from each other indicates tire creep, and this can be seen easily during a preflight inspection.

What are fusible tire plugs?

Fusible plugs offer protection from tire blowouts caused by thermal expansion that is generated in the tire under extra hard braking conditions. These fusible plugs are fitted in tubeless wheel hubs by means of a fusible alloy that melts under excessive heat conditions and allows the plug to be blown out by the tire air pressure. This prevents excessive pressure buildup in the tire by allowing the air to leak away slowly.

What are chimed tires?

A chimed tire has a special sidewall construction that takes the form of a ridge built onto the sidewall that diverts runway water to the side, reducing the amount of water thrown up into the intake of rear-mounted engines.

Describe a typical braking system.

Most aircraft braking systems use hydraulic fluid pressure to move friction brake pads (that are connected to small hydraulic pistons) against rotating brake plates to slow down the plates and therefore the wheel. The brake system is engaged either by the manual application of the brake pedals or by the automatic braking system that controls a brake metering valve for each wheel that adjusts the amount of piston movement and thereby the amount of pressure applied against each rotating wheel plate.

How does an automatic brake system decelerate an aircraft?

An automatic brake system regulates the amount of brake pressure by controlling the metering valve in the hydraulic brake line so as to maintain a constant deceleration rate until the aircraft reaches a complete stop. Brake application is regulated with the reverse thrust applied to maintain the selected deceleration rate.

Note: The autobrake system usually monitors the reverse thrust (where available and/or applied) and applies the autobrakes as a combined system to produce a constant deceleration rate to stop the aircraft. The reverse thrust is used at high speeds to decelerate the aircraft (where it is most efficient), and the brakes take over and are used at low speeds to bring the aircraft to a stop. Many modern aircraft also use ground spoilers to assist in stopping the aircraft on landing or during an aborted takeoff. (*See Q: What is the most effiective system for stopping at high speed? page 169.*)

Autobrake selection controls the magnitude of the deceleration rate. Typically, a system would have landing settings of 1, 2, 3, and maximum and a takeoff setting of RTO (rejected takeoff).

Tire temperature prior to takeoff depends on what factors?

Prior to the takeoff run, the following factors affect tire temperature:

1. Outside air temperature (OAT) (reflects the tire ambient temperature)
2. Aircraft weight
3. Taxi time
4. Amount of braking (which generates heat in the brakes that is transferred to the tires)

During the takeoff run, tire temperature depends on takeoff run (TOR) time (i.e., determined by head/tailwind component, runway slope, and pressure altitude). All generate a temperature rise in the tires.

Brake temperature prior to takeoff depends on what factors?

Prior to the takeoff run, the following factor affects the brake temperature:

1. Taxi time (distance) and the number of brake applications used

During the takeoff run, the following factors affect the brake temperature:

1. Aircraft takeoff weight
2. Pressure altitude
3. Outside air temperature (OAT)
4. Runway slope
5. Tail/headwind

All generate a temperature rise in the brakes.

Why is it important to monitor pneumatic tire temperatures prior to takeoff?

It is important to monitor tire temperature because an increase can lead to an increase in tire pressure, which causes the tire to expand, possibly too its blowout limit.

Why is it important to monitor brake temperature?

It is important to monitor the brake temperature, especially prior to commencing the takeoff run. This is done to ensure that the aircraft's inertia, which has to be dissipated through the wheel brakes, will not cause the brakes to overheat and lose efficiency, bind, and even cause tire failure or fire if the takeoff were abandoned at V_1. Therefore, brake temperature must be below a certain limit prior to commencement of the takeoff run to ensure that the maximum brake energy limit is not exceeded during an aborted takeoff run.

Which brakes get the hottest during a landing?

The downwind (with a crosswind) wheel brakes.

What is the most effective system for stopping at high speed?

(*See Chapter 9, "Flight Operations and Technique," page 287.*)

What is the purpose of antiskid systems?

The purpose of an antiskid system is to sense when the wheels are locked, i.e., not spinning, and to release brake pressure through the brake-modulating/metering valve system to generate wheel spin.

Wheel spin is required to maintain directional control and braking efficiency, which are necessary to ensure landing performance and are especially important in the operation of modern aircraft with high speeds, low drag, and high weights, particularly when coupled with operation from short runways in bad weather.

How do antiskid systems work?

First, the antiskid system has a detection system that senses the moment a wheel stops rotating, which it interprets as meaning that the wheel is off the ground (hydroplaning). Once it senses that the wheel is not rotating, the antiskid system releases brake pressure in the brake-metering/modulating valve system to release the brake pad from the brake wheel plate and therefore reinitiate wheel spin. (*See Q: How do you control an aquaplane? page 322.*)

What does an antiskid system protect against?

An antiskid system protects against the following:

1. *Locked wheel.* This can sometimes occur very rapidly, especially on a wet or icy runway with a lightly laden aircraft, due to excessive brake pressure for the prevailing conditions.
2. *Wheel skid/slip*
3. *Hydroplaning/aquaplaning*
4. *Touchdown.* An antiskid circuit maintains zero brake pressure prior to touchdown, which thus protects against locked wheels at touchdown.

How is aircraft cabin pressure measured?

Cabin pressure is measured as a differential between the ambient atmospheric air pressure outside the aircraft and the air pressure inside the cabin. This pressure measurement is known as *differential pressure,* e.g., 8.21 psi. The cabin pressure is normally greater than the external atmospheric pressure, and therefore, the cabin is often referred to as being blown up. The differential pressure value relates to the pounds per square inch force on the inside of the fuselage pushing outward.

How is the aircraft (cabin) pressurized/pressure controlled?

Aircraft (cabin) pressurization is controlled by the cabin's outflow air valve.

Conditioned air is pumped into the cabin as the air conditioning system dictates. However, the exhaustion of the air from the cabin is regulated by the main outflow valve to either build up, reduce, or to maintain an internal pressure as per the requirement of the pressure controller to a given differential pressure value. That is, on the ground with the outflow valve open, the internal cabin pressure is the same as the external atmospheric pressure; thus the differential pressure is 0 psi.

At what cabin altitude should a pilot go on oxygen?

10,000 ft.

What oxygen supply is delivered when selected to normal?

Normal selected on a pilot's oxygen mask delivers a sea-level pressure mixture of oxygen and ambient cabin air on demand.

Why and when should a pilot use 100% oxygen on demand?

100% oxygen on demand is used for life-support reasons to maintain the partial pressure of oxygen at or near sea-level pressure. Sea-level atmospheric pressure is 14.7 psi, and 21 percent of the atmosphere is oxygen; therefore, sea-level oxygen pressure is 3.08 psi.

It is not practical to keep the aircraft at the atmosphere's sea-level pressure of 14.7 psi, but it is practical to increase the percentage volume of oxygen, thereby maintaining sea-level oxygen partial pressure.

When do you use emergency 100% oxygen?

100% emergency oxygen is pure oxygen supplied continuously under positive pressure. It is used

1. For life-support conditions above 34,000 ft.
2. For medical conditions, especially suspected hypoxia
3. Whenever smoke and/or other harmful gases are present in the cabin

Note: Normal oxygen supply mixes oxygen with ambient air on demand. Therefore, any cabin fumes would be mixed and supplied to the crew if a normal oxygen supply were selected. Emergency 100% oxygen supplied under positive pressure does not mix ambient air and therefore ensures a nonharmful and constant oxygen supply.

What indications are there of a discharged crew oxygen bottle as a result of excess pressure?

First, a green disk on the aircraft's fuselage is blown out whenever excess pressure occurs in the crew's oxygen cylinder. Second, oxygen pressure gauges in the cockpit will indicate below normal or zero on most system types.

At what cabin altitude should passengers go on oxygen?

14,000 ft.

How is the passenger oxygen system activated?

The passenger oxygen system is activated either by the flight crew manually automatically by a barometric pressure controller that releases locking pins that allow the masks to drop from their overhead compartment whenever it senses a 14,000-ft cabin altitude.

When a passenger pulls on one of the masks (attached to an individual oxygen generator), an electrical firing mechanism mixes the chemical agents that generate the oxygen, which is then supplied continuously to all the attached masks until the system is emptied.

Prior to takeoff, what are the oxygen requirements for a flight?

For an individual flight, there must be enough oxygen in the crew system to allow for a 15- to 20-minute descent (i.e., cruise altitude down to 10,000 ft), plus enough oxygen for the rest of the flight to the nearest diversion airport based on a worst-case scenario.

In addition, there must be enough passenger oxygen for a 15- to 20-minute descent, plus enough oxygen for 10 percent of the required amount for the rest of the flight to the nearest diversion aerodrome based on a worst-case scenario.

What elements are required for a fire?

The elements required for a fire are (1) oxygen, (2) a combustible material (fuel), and (3) an ignition source (heat).

What is the most practical way to eliminate a fire?

Remove its oxygen supply.

What type of extinguisher should be used for electrical and flammable fires and normally found in the cockpit?

Bromochlorodifluoromethane (BCF). BCF is a liquefied gas agent that vaporizes on deployment.

What color container is BCF stored in?

Bromochlorodifluoromethane (BCF) is stored in signal red, purple-brown, or green containers.

What is the greatest contamination of fuel?

Water.

What steps can be taken to safeguard against water contamination?

1. Water drains in fuel tanks.

Note: Water is heavier than kerosene, and therefore, it falls to the bottoms of tanks, where it can be drained away.

2. Fuel heaters, typically used with gas turbine engines, are used to heat the fuel and evaporate the water prior to delivery to the engines.
3. Atmosphere exclusion in the fuel tanks.

Note: The main source of water contamination is the atmosphere that remains in the fuel tank. Thus keeping the tank full of fuel removes any air and thus prevents any water contamination.

What is specific gravity?

Specific gravity (SG) of a substance is the ratio of the weight of a unit volume of the substance to the weight of the same volume of water under the same conditions of temperature and pressure.

1 liter of fresh water weighs 1 kg = SG 1.0

1 liter of jet A1 fuel weighs 0.8 kg = SG 0.8

What affects the specific gravity of a substance?

The specific gravity of a substance varies with its density, which in turn varies inversely with its temperature. That is, as temperature increases, the SG decreases.

How is fuel measured?

The quantity of fuel on board an aircraft usually is measured in terms of its mass, i.e., in either kilograms (kg) or pounds (lb). By using the SG, you can calculate the mass of a given volume of fuel, and vice versa.

Convert fuel weights to volume, and vice versa.

Using a SG of 0.8,

$$\text{Weight in kilograms} \div 0.8 = \text{volume in liters}$$

$$\text{Volume of liters} \times 0.8 = \text{weight in kilograms}$$

Note: 1 liter = 3.7853 U.S. gallons

= 4.5459 imperial gallons

1 kilogram = 2.2046 lb

Why is fuel measured in terms of a mass (weight) rather than by volume?

Fuel is not measured in terms of a volume because a volume of fuel will increase with a temperature rise and, more important, will decrease with a temperature drop. However, with a rise in temperature, the specific gravity decreases, and with a drop in temperature, the specific gravity increases, which ensures that the indicated mass (weight) of fuel remains the same.

Obviously, if fuel were uploaded in terms of a volume under high-temperature conditions and a change in temperature reduced the volume of fuel before or during a flight, then a situation could exist whereby the aircraft had insufficient fuel to complete the journey.

What precautions should be taken before and during aircraft fueling?

The following procedures should be carried out:

1. Before fueling commences, fueling zones should be established. Within these zones, the following restrictions should be enforced:
   a. No smoking.
   b. If the auxiliary power unit (APU) exhaust discharges into the zone, then it should not be started at any time during refueling to protect against the APU exhaust igniting the fuel vapors.

Note: If the APU is on prior to refueling, then it is safe to leave the APU running during refueling.

   c. Ground power units should be located as far away as practical from the fueling zones and not connected or disconnected while fueling is in progress to protect against any electrical arching igniting fuel vapors.
   d. Fire extinguishers should be located so as to be readily accessible.

2. During fueling, the aircraft should be earthed and bonded to the fueling equipment.

Note: For overwing gravity refueling, the hose nozzle should be bonded to the aircraft structure before removing the tank filler cap. For pressure refueling, the tank pressure relief valves, where fitted, should be checked if possible, and the fuel hose bonding lead should be connected before attaching the fueling nozzle.

3. Passengers embarking and disembarking the aircraft during fueling operations should do so under supervision, and their route should avoid the refueling zones.

What is a deicing system?

A deicing system is one in which ice is allowed to build up on a surface and is then removed, e.g., with pneumatic leading-edge boots.

What is an anti-icing system?

An anti-icing system is one in which ice is prevented from building up on a surface, e.g., thermal or electrical anti-icing systems on engine cowls.

What is the purpose of cockpit window heating?

The purpose of heating the windows is to prevent them from breaking as a result of bird strikes. For example, if the window freezes, a bird strike is likely to crack the window. However, if the window is heated, it is more flexible and therefore able to absorb a bird strike without cracking or breaking.

Aircraft Electrical Systems

Technical questions asked in an interview about aircraft electrical systems tend to be type-specific. Thus knowledge of your particular aircraft is required, and as such, only general questions are covered in this section.

What is electricity?

Electricity is the movement of electrons (electromotive force and current) that produces a power.

What is a volt a measure of?

A volt is a unit of electrical force/pressure (expressed as a V). Volts are a measure of

1. Electromotive force (EMF), i.e., the electrical pressure available from a source of electrical energy used to produce electron flow (current).
2. Potential differences (pd) in the EMF level between any two circuit points that creates an electrical pressure, which can produce electron flow from the less positive to the more positive circuit point.

Note: 1 kilovolt (1 kV) = 1000 V

1 millivolt (1 mV) = 1/1000 V (0.001 V)

What does a voltmeter (gauge) show?

A voltmeter indicates the number of volts produced by an electrical source, e.g., a generator, or the potential difference in a system between any two points, e.g., a transformer.

What is an amp (ampere) a measure of?

An amp (ampere) is a unit of electric current (expressed as an I), i.e., the number of electrons flowing in a circuit when an electrical pressure/force (volts) is applied.

Note: 1 milliamp (1 mA) = 1/1000 A (0.001 A)

What does an ammeter (gauge) show?

An ammeter indicates the number of amps/ampere (quantity of electrons flowing in a current) in an electric circuit.

Note: A quantity of amps is often referred to as a current.

What is an ohm a measure of?

An ohm is a unit of electrical resistance (expressed by the symbol R), i.e., the degree of resistance or opposition to current flow (or amps).

Note: 1 kilohm (1 kΩ) = 1000 Ω

1 megaohm (1 MΩ) = 1,000,000 Ω

What does an ohmmeter (gauge) show?

An ohmmeter measures and indicates the amount of ohms (electrical resistance) to the electric current.

What is a watt a measure of?

A watt is a unit of electrical power (expressed by the symbol P), where the electrical power (watt) is a product of the volts and amps being used in the electrical circuit. That is,

$$\text{Watts} = \text{volts} \times \text{amps}$$

Note: 1 kilowatt (1 kW) = 1000 W

1 horsepower (1 hp) = 746 W (modern engines are measured by horsepower)

What does a wattmeter (gauge) show?

A wattmeter measures and indicates the number of watts (electrical power) being consumed.

What is a series circuit?

A series circuit is one in which there are no junctions or branches, just one circuit path.

The word *series* is also applied to the way electrical sources, resistors, or loads are situated (i.e., in a chainlike connection through which the same current passes), and their effect on the electric current is additional, namely, connected in series.

What is a parallel circuit?

A parallel circuit is one in which there are two or more alternative paths that subsequently reunite. A feature of this arrangement is that the total current is shared between the parallel circuit paths at its first junction and then restored to a single value at a second junction.

The word *parallel* is also applied to the way electrical sources, resistors, and loads are connected to different circuit paths (branches). In this way, the failure of one device is somewhat isolated and does not affect them all.

What is the purpose of a fuse?

A fuse is a piece of wire with a low melting point that is placed in series with the electrical load and either melts, blows, or ruptures when a current with a higher value than its ampere rating is placed on it, thus protecting the electrical load/equipment from excessive power surges.

How many spare fuses should be carried on an aircraft?

There should be a minimum of 10 percent (or at least three) spare fuses of the total number of each rated fused installed.

What are circuit breakers?

Circuit breakers (CBs) are thermal devices placed in series with an electrical load. They open circuit and thus cut off the electrical equipment when they experience an abnormal (overload) current operating condition.

Open-circuit conditions are indicated by a CB reset push button being visible on a CB panel. Pressing the CB button in will reengage and complete the circuit.

What are non-trip-free circuit breakers?

A non-trip-free circuit breaker can be held in under a fault condition as an emergency measure to complete the circuit and engage the electrical load.

What is a trip-free circuit breaker?

A trip-free circuit breaker will not make any internal contact by pressing the reset button with an overload condition in the electric circuit.

Note: This type of circuit breaker cannot be reset by holding in the reset button under a fault condition.

What is dc electrical power?

Direct current (dc) electrical power flows in only one direction around a circuit and has no appreciable variation in its amplitude.

What are an aircraft's typical sources of dc electrical power?

An aircraft's typical sources of dc electrical power are

1. Dc generator
2. Battery
3. Ground dc supply
4. Rectifier (changes ac to dc)

A transformer-rectifier unit (TRU) changes ac to dc and changes its voltage level.

How does a battery create electrical energy?

A battery uses chemical action to separate electrons from their parent atoms and thus generate dc electricity. The material contained in battery cells determines the voltage output, whereas the amount of current depends on the size of the plates that store the electric current.

What is a primary cell battery?

A primary cell battery uses up the chemicals that produce the electrical energy. Eventually, the cell runs out of one or more of the chemicals, and the battery is said to be *discharged.*

What is a secondary cell battery?

A secondary cell battery can reverse the chemical changes that take place in its discharge. The secondary cell can convert electrical energy back into chemical energy, which is known as *charging.* This reconverted chemical energy can then be used to create electrical energy. By careful handling, a secondary cell can be charged and discharged many times.

What is alternating current?

Alternating current (ac) continuously reverses its direction of flow in an electric circuit. The complete reversal of current flow is known as a *cycle* and occurs very rapidly. For example, aircraft ac electrical systems reverse the current flow 400 cycles per second.

Note: 1 cycle per second is called a hertz.

What advantages does ac electrical power have over dc electrical power?

The advantages of ac power are as follows:

1. Ac generators (alternators) are simpler and more robust in construction than the dc machines.
2. The power-to-weight ratio of ac machines is much better than that of comparable dc machines.
3. The supply voltage can be converted to a higher or lower value with great efficiency using transformers.
4. Any required dc voltage can be obtained simply and efficiently by using transformer-rectifier units (TRUs).
5. Three-phase ac motors can be operated from a constant frequency source (alternator).
6. Ac generators do not suffer from commutation problems associated with dc machines and consequently are more reliable, especially at high altitudes.
7. High-voltage ac systems require less cable weight than comparable power low-voltage dc systems.

What are an aircraft's typical sources of ac power?

The typical sources of ac electrical power for an aircraft are as follows:

1. Ac generators (alternators), either engine, APU, or ram air turbine (RAT) driven.
2. Inverters (converts dc to ac). Static inverters produce constant frequency ac power. Rotary inverters produce power for frequency-wild ac systems.
3. Ground ac supply.
4. Transformers (change the ac voltage level).

What is a transformer?

A transformer is an electrical device without any moving parts that uses magnetic induction between two windings, known as the *primary* (input) and *secondary* (output) *coils,* to increase or decrease the alternating current.

Note: When the secondary coil has more turns than the primary, the voltage is increased, or stepped up. When the secondary coil has fewer turns than the primary, the voltage is decreased, or stepped down.

Transformers are used in ac electrical systems to step up or down the voltage to meet the requirement of the load on it.

Note: Often transformers are paired with rectifiers, namely, as a transformer-rectifier unit (TRU), to convert the current to dc as well as the voltage level.

What do constant-speed drive units (CSD) achieve?

Constant-speed drive units (also known as a *generator drive*) maintain the ac frequency output of an alternator, normally, to 400 Hz.

The number of cycles per second is the frequency of the ac supply. That is, a cycle is the complete reversal of direction of the electric ac current, and one cycle per second is measured as a hertz.

How does a basic constant-speed drive (CSD) system work?

The basic constant-speed drive unit (also known as a *generator drive*) consists of an engine-driven hydraulic pump that drives a hydraulic motor, which itself drives the alternator. Most CSD units are capable of maintaining the alternator output frequency to within 5 percent of 400 Hz.

The CSD unit can be disconnected from the engine input drive in the unlikely event of a malfunction. This allows both the drive unit and the alternator to become stationary, thus eliminating any chance of a malfunction affecting the engine.

When can you disconnect and then reconnect a CSD unit?

The constant-speed drive (CSD) unit can be disconnected at any time, but reconnection can only be done manually on the ground following shutdown of the engine.

What are the effects of a buildup of static electricity on an aircraft?

Static electricity has the potential to create a spark, inducing a fire risk and/or creating radio interference, if it is allowed to build up in a section of an aircraft.

How is the risk of static electricity reduced?

Bonding prevents any part of the aircraft from building up static electricity.

Bonding is the joining by flexible wire strips of each part of the aircraft's metal structure or metal components to each other, thus providing an easy path for the electrons to move from one part of the aircraft to another. Static wicks are fitted to the trailing of the aircraft surfaces to dispense static electricity into the atmosphere, thus reducing the risk of flash fires and radio interference.

What are the basic parameters of an aircraft's electrical system?

The basic parameters of an aircraft's electrical system are

1. No paralleling of the ac sources of power.
2. All generator bus sources have to be manually connected through the movement of a switch that also will disconnect any previously existing source.

Chapter

7

Performance and Flight Planning

Loading

What are the following aircraft weight definitions?

Basic weight. The weight of the empty aircraft with all its basic equipment plus a declared quantity of unusable fuel and oil.

Note: For turbine-engined aircraft and aircraft not exceeding 5700 kg, the maximum authorized basic weight may include the weight of its usable oil.

Variable load. The weight of the crew, crew baggage, and removable units, i.e., catering loads, etc.

Variable load = APS − basic weight

APS weight. Aircraft prepared for service (APS).

APS = basic weight + variable load

Payload. Passengers and/or cargo.

Disposable load. The weight of the payload and fuel.

Disposable load = TOW − APS

Ramp weight. Ramp weight (RW) is the gross aircraft weight prior to taxi.

Note: RW must be within its structural maximum (certificate of airworthiness) weight limit.

RW = TOW + fuel for start and taxi

MTOW. Maximum takeoff weight (MTOW) is the maximum gross weight of the aircraft permitted for takeoff.

Note: Sometimes a performance-limited MTOW (i.e., short runway, obstacle clearance) may limit the aircraft to a weight less than its structural maximum (certificate of airworthiness) weight.

MLW. Maximum landing weight (MLW) is the maximum gross weight of the aircraft permitted for landing.

Note: Sometimes a performance-limited MLW (i.e., short runway, obstacle clearance) may limit the aircraft to a weight less than its structural maximum (certificate of airworthiness) weight.

ZFW. Zero fuel weight (ZFW) is a wing loading structural maximum weight.

$$\text{ZFW} = \text{payload} + \text{APS}$$

$$\text{ZFW} = \text{MTOW} - \text{fuel weight}$$

Thus the maximum ZFW determines the maximum permissible payload.

For weight and center of gravity questions, see Chapter 1, "Aerodynamics," subchapter, "Weight and Aircraft Momentum," page 11.

What factors determine the loading (weight and balance) of an aircraft?

The factors that determine the loading (weight and balance) of an aircraft are

1. To ensure that the following combined component weights do not exceed the aircraft's overall gross weight limitations, i.e., MTOW, ZFW, structural maximum (certificate of airworthiness) aircraft weight, etc.:
   a. Cargo
   b. Baggage
   c. Crew and passengers, including personal effects (approximate weights can be used)
   d. Removable equipment
   e. Fuel (SG weight)
2. The distribution of the weights ensures that the center of gravity is within its limits to longitudinally balance the aircraft. (*See Q: Describe center of gravity, page 12.*) This is accomplished by adding up the various component moments to obtain the total moment. (*See Q: Describe center of gravity moment, page 12.*)

The various component weights act at known positions relative to the fixed datum (arm). Each of these has a moment, which we can calculate by multiplying the individual weight by its arm from the datum. We obtain the total moment by adding up all the moments of these component weights and then by using the formula

$$\text{Total arm} = \frac{\text{total moment}}{\text{total weight}}$$

we can calculate the center of gravity position.

Carrying out the weight and balance/loading calculation for an aircraft is essential for a safe flight.

In loading an aircraft to obtain maximum range, would you load it with an aft or forward center of gravity?

(*See Chapter 1, "Aerodynamics," page 16.*)

Takeoff and Climb

What is the takeoff run available (TORA)?

Takeoff run available (for all engine operations) is the usable length of the runway available that is suitable for the ground run of an aircraft taking off. In most cases this corresponds to the physical length of the runway.

What is the takeoff run required (TORR)?

Takeoff run required (for all engine operations) is the measured run (length) required to the unstick speed (V_R) plus one-third of the airborne distance between the unstick and the screen height. The whole distance is then factored by a safety margin, usually 15 percent.

What is the runway clearway?

The clearway is the length of an obstacle-free area at the end of the runway in the direction of the takeoff, with a minimum dimension of 75 m either side of the extended runway centerline that is under the control of the licensed authority.

Note: The clearway surface is not defined and could be water. It is an area over which an aircraft may make a portion of its initial climb to a specified height, i.e., to the screen height, 35 ft.

What is the takeoff distance available (TODA)?

Takeoff distance available (for all engine operations) is the length of the usable runway available plus the length of the clearway available, within which the aircraft initiates a transition to climbing flight and attains a screen height at a speed not less than the takeoff safety speed (TOSS) or V_2.

$$\text{TODA} = \text{usable runway} + \text{clearway}$$

TODA is not to exceed 1.5 × TORA (takeoff run available). The TODA/R and TOSS/V_2 are the vital performance figures for a pilot during the takeoff.

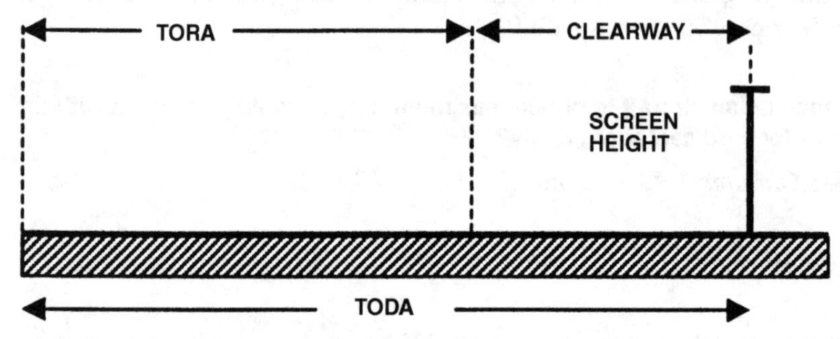

Figure 7.1 Takeoff distance available.

What is the takeoff distance required (TODR)?

Takeoff distance required (for all engine operations) is the measured distance required to accelerate to the rotation speed (V_R) and thereafter effect a transition to a climbing flight and attain a screen height at a speed not less than the takeoff safety speed (TOSS) or V_2. The whole distance is factored by a safety margin, usually 15 percent.

What is screen height?

Screen height relates to the minimum height achieved over the runway before the end of the clearway should an engine failure occur on takeoff. The screen height also marks the end of the takeoff distance.

How high is the screen height for propeller and jet aircraft?

The screen height for propeller-engined aircraft in dry conditions is 50 ft.

Note: Most propeller aircraft have an increased accelerate/stop distance in wet conditions but no change in V_1 or screen height.
The screen height for jet aircraft in dry conditions is 35 ft.

Note: Less than propeller aircraft due to the lower C_L of the jet aircraft's swept wing. In wet conditions, the jet aircraft's screen height is reduced to a minimum of 15 ft in most cases. This is so because when an engine failure occurs at the worst point, i.e., after V_1 (wet or dry) and prior to V_R, a proportion of the airborne distance is added to the ground run. (*See Q: How does screen height change with a wet V_1? page 191.*)

Note: V_2 will only be achieved at 35 ft; therefore, at a reduced screen height of 15 ft the aircraft speed will be less than V_2.

Thus screen height relates to engine failure scenarios and changes with runway conditions for jet aircraft, i.e., 35 ft for 1 engine inoperative/dry conditions and 15 ft for 1 engine inoperative/wet conditions.

What is the runway stopway?

The stopway is the length of an unprepared surface at the end of the runway in the direction of the takeoff that is capable of supporting an aircraft if the aircraft has to be stopped during a takeoff run.

What is the emergency distance available (EMDA)/accelerate stop distance available (ASDA)?

Emergency distance available (also known as the *accelerate stop distance available,* or ASDA) is the length of the takeoff run available, usually the physical length of the runway, plus the length of any stopway available. That is,

$$\text{EMDA/ASDA} = \text{usable runway} + \text{stopway available}$$

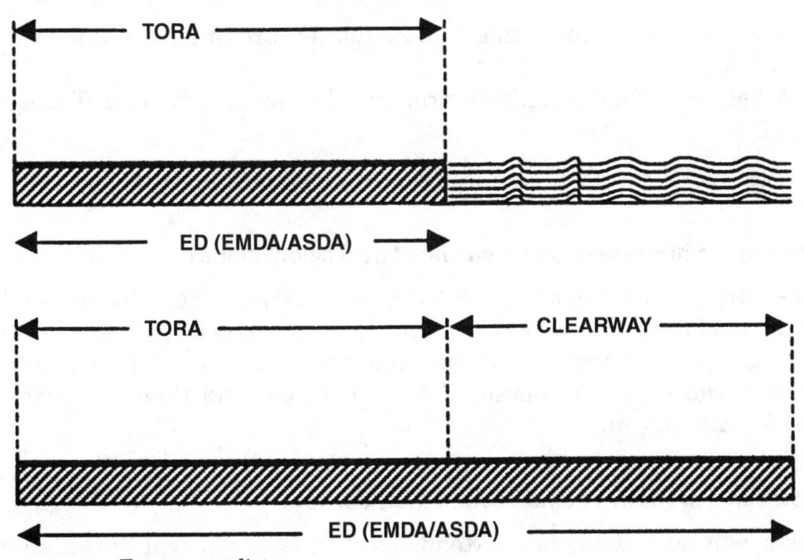

Figure 7.2 Emergency distance.

What is the emergency distance/required (ED/EDR)?

Emergency distance required is the distance required to accelerate during the takeoff run on all engines to the critical speed, V_1, at which point an engine failure is assumed to have occurred, and the pilot aborts the takeoff and brings the aircraft to a halt before the end of the runway or stopway if present; i.e., RTO.

The whole emergency distance is factored by a safety margin, normally 10 percent.

Note: The use of reverse thrust in the EDR calculation differs from authority to authority but usually it is not factored in the EDR calculation.

ED is sometimes referred to as an accelerated stop distance. The EDR must not exceed the EMDA.

Explain balanced and unbalanced fields.

A balanced field exists when TODA = EMDA(ASDA) or, in other words, when the end of the clearway is the end of the stopway, and the aircraft achieves the screen height over the end of the runway in all cases.

Note: A balanced field may be assumed to exist if that part of the clearway which extends beyond the stopway is ignored; therefore, the lower the takeoff distance available (TODA) or emergency distance available (EMDA), the more balanced is field length.

Thus an unbalanced field exists when TODA is greater than EMDA.

Note: TORA (runway length) does not feature in balanced field calculations.

A balanced field length determines the maximum takeoff weight (MTOW).

Note: Airbus calls a balanced field a quick reference table.

What is the purpose of using balanced field calculations?

The purpose of using a balanced field calculation is to optimize the V_2 climb performance (second segment) with a correct V_1/V_R speed from a single performance calculation/chart without having to perform a second and separate increased V_2 calculation and then readjusting the V_R calculation.

How can a stopway extend beyond the clearway?

A stopway sometimes may extend beyond the clearway if the length of the clearway is limited because of an obstruction within 75 m of the runway/stopway centerline. (*See Q: What is the runway clearway? page 185.*) However, this obstruction does not limit the stopway, which only needs to be as wide as the runway.

What is the significance of the 40- to 100-knot call during the takeoff roll?

The 40- to 100-knot call during the takeoff roll is used to check the requirements that need to be established by the called speed. These requirements include

1. Directional control surface (vertical tailplane) starts to become effective with all engines operating.
2. Takeoff engine pressure ratio (EPR) should be set by this check speed so that the pilot is not chasing engine needles for a prolonged period during the takeoff roll.
3. Cross-check the airspeed indicator gauges to ensure their accuracy and reliability.

In addition, type-specific requirements also might need to be established by the takeoff roll check speed.

What is VMU speed?

This is the minimum demonstrated unstick speed at which it is possible to get airborne on all engines and to climb out without hazard.

What is the critical speed?

Critical speed is the lowest possible speed on a multiengine aircraft at a constant power setting and configuration at which the pilot is able to maintain a constant heading after failure of an off-center engine. VMCG/A/L are particular configurations and stage of flight critical speeds.

What is VMCG speed?

VMCG is the minimum control speed on the ground for a multiengine aircraft at a constant power setting and configuration, at and above which it is possible to maintain directional control of the aircraft around the normal/vertical axis by use of the rudder to maintain runway heading after failure of an off-center engine. (*See Q: How would you teach a student about VMCG/A? page 192.*)

What is V_1 speed?

V_1 is the decision speed in the event of an engine failure during the takeoff roll, at which it is possible to continue the takeoff and achieve the screen height (*see Q: What is screen height? page 186*) within the normal takeoff distance available or to bring the aircraft to a full stop within the emergency distance available (accelerate stop distance).

The takeoff must be abandoned with an engine failure below V_1, and the takeoff must be continued with an engine failure above V_1.

Note: If the TOW is limited by TODA, TORA, or EMDA, the V_1 speed relates to a single point along the runway where the pilot will have the decision to continue or abort the takeoff in the event of an engine failure.

V_1 cannot be less than VMCG; V_1 cannot be greater than V_R or VMBE.

What is a range of V_1 speeds?

When the planned takeoff weight (TOW) is not field-length-limited, i.e., not limited by TORR, TODR, or EMDR, there may be a range of V_1 speeds; e.g., between 125 and 135 knots when the pilot has a stop-go option in the event of an engine failure. The minimum V_1 speed in the range is still restricted by the VMCG speed, and the maximum V_1 is still restricted by the VMBE.

A range allows the V_1 speeds to be raised or lowered to achieve a different departure profile or climb technique, i.e., an increased TOW or an increased V_2 climb for obstacle clearance. However, normally, a single V_1 speed is chosen within the range prior to commencing the takeoff run. Thus, if an engine failure does occur, the decision effectively has already been determined, thus removing any delayed response and indecision that an analysis of a V_1 range at the time of the engine failure might precipitate, which improves operational safety.

How does weight affect the V_1 speed?

If the field length is limiting, the greater the aircraft weight, the lower is the V_1 speed. This means that the lower V_1 speed provides a greater stopping distance while ensuring that V_1 remains greater than VMCG and VMU.

If the field length is not limiting, the greater the aircraft weight, the higher is the V_1 speed, providing V_1 remains less than the VMBE speed and the field length emergency stopping distance is not compromised.

What is the difference between a dry V_1 and wet V_1?

A dry (maximum) V_1, is the normal decision speed that following an engine failure allows the takeoff to be continued safely within the TODA or to be stopped safely within the EMDA. A wet (minimum) V_1 is the maximum speed for abandoning a takeoff on a contaminated runway. A wet V_1 improves the stopping capabilities (final stop point) back to the dry conditions level but degrades the takeoff chances with a reduced screen height in the event of a takeoff being continued. A recommended wet V_1 for contaminated conditions is the dry V_1 − 10 knots. Thus wet V_1 is a lower speed than dry V_1.

The wet V_1 is not a V_1 speed because it does not imply any ability to continue the takeoff following an engine failure, and unlike a dry V_1, this speed may be less VMCG. Therefore, a takeoff from a wet runway may result in a risk period between the maximum speed for abandoning the takeoff (wet V_1) and the normal V_1 speed, during which, in the event of an engine failure, the speed is too high for a successful stop on a contaminated runway.

How does a contaminated runway (ice and rain) affect distance and V_1 speed?

For a given aircraft weight on a contaminated runway, the emergency distance required is increased because of a reduced braking ability. Also, a contaminated runway has a slower acceleration, and therefore, the TORR is increased, which limits the stopping distance available if the takeoff is field-length-limited. This effect is normally built into takeoff run performance graphs.

If distance is limited, the normal dry V_1 offers the best compromise in risk associated with a continued or aborted takeoff. Namely, there is a risk of not being able to stop within the EMDA. However, it is a remote possibility that an engine failure will occur at the worst point, i.e., after the wet V_1 and before the dry V_1; therefore, an incident in the transition to flight in this scenario is a low risk. Furthermore, even if you lose an engine between the wet and dry V_1 speeds, aborting the takeoff probably will only result in a runway overrun at a low speed.

If there is a distance to spare, i.e., not field length limited, then a progressive reduction in the V_1 speed to the wet V_1 reduces the risk associated with the aborted takeoff without unduly compromising the continued takeoff. This is so because of the increased airborne length of the effective clearway over the runway. (*See Q: What is the difference between a dry V_1 and a wet V_1? page 190.*)

How does screen height change with a wet V_1?

Screen height is reduced for a jet aircraft using a wet V_1. This is due to a portion of the airborne distance being added to the ground run as a result of the increased ground run used between the wet V_1 and V_R if an engine failure occurs at the worst point, i.e., just after the wet V_1 and prior to V_R.

Note: However, most propeller aircraft have no change in V_1 and screen height in wet conditions.

What is VMBE speed?

Maximum brake energy speed (VMBE) is the maximum speed on the ground from which a stop can be accomplished within the energy capabilities of the brakes. (*See Q: Describe brake energy limits, page 203.*)

What do you do if V_1 is greater than VMBE?

If the V_1 speed exceeds the maximum brake energy speed (VMBE), then the aircraft's takeoff weight has to be reduced until the V_1 speed is less than or equal to VMBE to ensure that the aircraft does not exceed its brake energy limit. (*See Q: Describe brake energy limits, page 203.*) Hence VMBE can limit V_1 and thus MTOW, especially on downward-sloping runways with a tailwind.

An aircraft will have a set weight reduction for each knot of speed.

Note: V_R and V_2 need to be redetermined for the lower aircraft weight.

What is V_R speed?

V_R (rotation speed) is the speed at which the pilot initiates rotation during the takeoff to achieve V_2 at the screen height, even with an engine failure. V_R cannot be less than 1.05 VMCA/1.1 or 1.05 VMU.

What is the relationship of V_1 and V_R?

V_R is either greater or equal to V_1 but never less than V_1.

What is V_S speed?

V_S (stall speed) is the speed at which the airflow over the wings will stall. The stall speed varies with aircraft weight and configuration. The stall speed is the reference speed for the other performance speeds, i.e., V_2, V_{ref}, etc.

What is V_a speed?

Maneuvering speed.

Maneuvering speed is the airspeed at which maximum elevator deflection causes the stall to occur at the airframe's load factor limit. V_a for maximum aircraft weight is specified in the flight manual.

What is VMCA speed?

VMCA is the minimum control speed in the air for a multiengine aircraft in the takeoff and climb-out configuration, at and above which it is possible to maintain directional control of the aircraft around the normal/vertical axis by use of the rudder within defined limits after the failure of an off-center engine. (*See Q: How would you teach a student about VMCG/A? page 192.*)

How would you teach a student about VMCG/A?

VMCG/A relates to the minimal directional (heading) control speed on the ground or in the air, at which the turning moment produced by the vertical tailplane with maximum rudder deflection is sufficient to balance the yawing moment of the aircraft nose when the aircraft loses an off-center engine (asymmetrical power).

The heading of an aircraft is determined by the direction of the nose of the aircraft, which is pivoted about the normal/vertical axis at the center of gravity point. With an off-center engine failure, and assuming a constant power and configuration setting, the aircraft will yaw about the center of gravity point to the dead engine due to the asymmetrical thrust properties. This yaw changes the aircraft's direction. The magnitude of the yaw is a function of the asymmetrical thrust × aircraft nose to center of gravity arm. That is,

Yawing moment = asymmetrical thrust × nose to center of gravity arm

Whenever the aircraft is committed on the takeoff run, i.e., past V_1, or in the air, this yawing moment has to be balanced to maintain directional control of the aircraft. The directional control is provided by the aircraft's vertical tailplane and its rudder control surface, which is used to produce a turning moment to oppose the yawing moment of the aircraft. The vertical tailplane produces a turning moment that is based on the design of the surface, i.e., size (and assumes a maximum rudder deflection), and is a product of the vertical tailplane to center of gravity arm and the weight on the tailplane; where the weight is the air load force experienced. That is,

Turning moment = rudder to center of gravity arm × weight (air load force)

Because the rudder to center of gravity arm is a constant during a particular takeoff run or phase of flight, the vertical tailplane turning moment is influenced/determined by the weight on the vertical tailplane. The weight, as stated previously, is the force experienced, which is a product of

1. The angle of deflection (We always assume a maximum rudder deflection.)
2. Air density (has a minimal influence)
3. Airspeed

Therefore, the speed determines the aerodynamic force/weight over the vertical tailplane. The greater the speed, the greater is the rudder turning moment and the greater is the directional control for a given atmospheric condition, i.e., air density. Therefore, weight is a product of airspeed, and the formula can be expressed as

Turning moment = rudder to center of gravity arm × speed (VMCG/A)

In essence, the magnitude of the two opposing moments, i.e., the yawing moment of the aircraft's nose and the turning moment of the vertical tailplane, produces a seesaw effect. When the aircraft's yawing moment is dominant, the aircraft cannot maintain directional control, which results in the aircraft yawing off heading. When the vertical tailplane turning moment is dominant, the aircraft can maintain directional control, which results in the aircraft being able to maintain heading. Which moment is dominant depends on the airspeed; therefore, VMCG/A relates to the minimum control speed on the ground or in the air, at and above which the vertical tailplane turning moment is dominant over the yawing moment, and therefore, directional control of the aircraft is guaranteed.

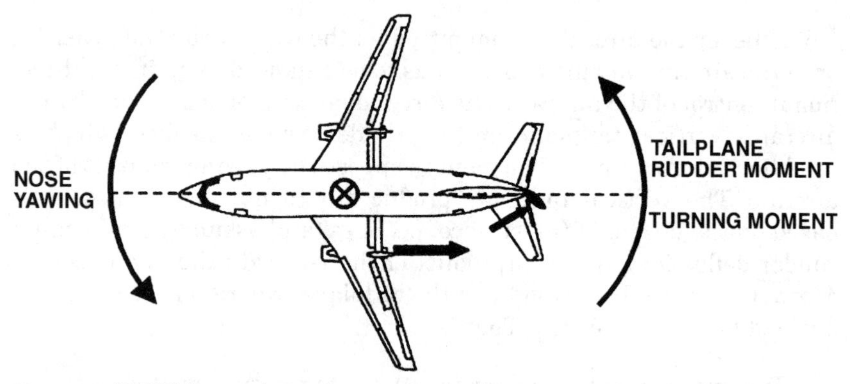

Figure 7.3 VMCG/A—yawing and turning moments.

When loading an aircraft, the position of the center of gravity is determined, and the center of gravity position affects the size of both the yawing and turning moments. Therefore, VMCG/A speeds vary with the center of gravity position. (*See Q: How does VMCG/A vary with center of gravity position? page 195.*)

To conclude, remember that VMCG/A relates to asymmetrical power configurations and guarantees directional control of the aircraft (not climb capabilities) whenever the aircraft's airspeed is equal to or greater than VMCG/A. During the ground takeoff run (TOR), if you lose an off-center engine below VMCG, you will not have directional control of the aircraft; therefore, you have to remove the asymmetrical thrust by closing the engines to regain directional control, and obviously, this results in an aborted takeoff. To allow for this action, VMCG always must be less than or equal to V_1, thus ensuring that you retain directional control at and above V_1. In the air, VMCA has to be equal or less than V_2 to ensure that the aircraft can always retain directional control.

How does VMCG/A vary with center of gravity position?

An aft center of gravity position requires a higher VMCG/A. (*See Qs: What is VMCG/A speed? pages 189 and 192; How would you teach a student about VMCG/A? page 192.*)

The turning moment acts around the center of gravity, and if the center of gravity is in the aft position, the vertical tailplane (rudder) moment arm will be shorter, and therefore, the vertical tailplane turning moment is less for a given airspeed. Thus the aircraft requires a higher minimum control speed (VMCG/A) with an aft center of gravity position.

Turning moment = rudder to center of gravity arm × speed (VMCG/A)

Conversely, the opposite is true for a forward center of gravity position. A forward center of gravity will have a longer arm, and therefore, the vertical tailplane turning moment is greater for a given speed, and thus the aircraft can have a lower VMCG/A.

If VMCG is limiting for the weight of the aircraft, what can you do?

Reduce takeoff thrust.

The vertical tailplane (rudder) turning moment is used to oppose/balance the asymmetrical thrust yawing moment to maintain directional control. Therefore, by reducing thrust, any off-center engine loss during the takeoff run has a reduced asymmetrical thrust imbalance, which thereby reduces the yawing moment experienced and thus requires a reduced tailplane turning moment to maintain directional control. And because the magnitude of the tailplane turning moment is a product of airspeed, a lower (VMCG) ground speed maintains the aircraft's directional control. Thus, reduced takeoff thrust gives rise to a lower VMCG.

What is the relationship between VMCG and V_1?

VMCG has to be equal to or less than V_1, thus ensuring that the aircraft can maintain directional control with an off-center engine failure at or above V_1, when the aircraft is committed to the takeoff and directional control of the aircraft is essential for safe operation.

If you had an engine failure between V_1 and V_R and you had a maximum crosswind, which engine would be the best to lose, i.e., upwind or downwind engine?

Upwind engine. This is so because the crosswind would then oppose the yawing moment of the downwind engine. (*See Q: How does a crosswind affect the critical engine? page 60.*)

What is V_2 speed?

V_2 speed is the takeoff safety speed achieved by the screen height in the event of an engine failure that maintains adequate directional control and climb performance properties of the aircraft.

Note: V_2 is also known as the *takeoff safety speed* (TOSS). V_2 cannot be less than $V_S \times 1.20$ and VMCA $\times$ 1.10.

What is the relationship between V_S and V_2?

V_2 is equal to or greater than $1.2 \times V_S$.

What is the difference between VMCA and V_2?

VMCA must be less than V_2. Normally, V_2 is equal to or greater than 1.1 $\times$ VMCA.

VMCA relates to the airborne directional control of the aircraft in the event of an off-center engine failure. V_2 relates to the directional control and a minimum climb performance of the aircraft in the event of an engine failure.

What is V_3 speed?

V_3 speed is the all-engine-operating takeoff climb speed the aircraft will achieve at the screen height.

What is V_4 speed?

V_4 speed is the all-engine-operating takeoff climb speed the aircraft will achieve by 400 ft, and is used as the lowest height where acceleration to flap retraction speed is initiated.

What are the main variables (conditions) that affect an aircraft's takeoff and landing performance?

An aircraft's takeoff and landing performance is subject to many variable conditions, including

1. Aircraft weight
2. Aircraft flap setting
3. Aerodrome pressure altitude
4. Air density/density altitude (temperature and pressure altitude)
5. Humidity
6. Wind
7. Runway length, slope, and surface (including wet or icy conditions)

How does aircraft weight affect takeoff performance?

Increased aircraft weight results in a greater takeoff distance required (TODR) and a reduced net takeoff climb gradient.

How does the use of flaps affect the aircraft's takeoff performance?

The effect of flaps on the takeoff performance, i.e., TOR/TOD, and climb performance varies between different aircraft types, especially between swept-wing (jet) and straight-wing (turboprop) aircraft and further with the degree of flap deployed on individual aircraft.

Swept-wing (jet) aircraft. Swept-wing aircraft require a low flap setting, i.e., takeoff flap, to improve the C_L during the takeoff. This has two positive effects. First, a low flap setting reduces the aircraft's takeoff run required (TORR) because the higher C_L lowers the stalling speed (V_S), which in turn reduces the V_2 and V_R speeds and results in the aircraft reaching its liftoff speed from a shorter ground takeoff run (TORR). Second, a low flap setting reduces the aircraft's takeoff distance required (TODR) because the increased C_L benefits outweigh the increased airborne drag and thereby improve the climb performance of the aircraft, resulting in the aircraft reaching the screen height over a shorter distance.

Maximum takeoff flaps may be used to reduce the ground takeoff run required when the field length is limiting or the runway surface is poor. However, airborne climb performance may be compromised due to an increase in drag, which reduces the lift-drag ratio and results in a reduced rate of climb (climb gradient) performance. (*See Q: How do flaps affect takeoff ground run? page 41.*)

The use of flap settings outside the takeoff range would increase aerodynamic drag during the ground run, causing a slower acceleration that results in an increased TORR and then once airborne would significantly degrade the climb gradient performance because of the poor lift-drag ratio that results in an unacceptable increase in TODR.

Straight-wing (turboprop) aircraft. Straight-wing (turboprop) aircraft usually require a low flap setting to improve the C_L during takeoff. This has the effect of reducing the aircraft's takeoff run required (TORR) because the higher C_L lowers stalling speed (V_S), which in turn reduces V_R speed and results in the aircraft reaching its liftoff speed (V_R) from a shorter ground run (TORR). However, the takeoff distance required (TODR) to the screen height may not be reduced significantly, if at all, because, as well as increasing lift, flap deployment also increases drag, thus reducing the lift-drag ratio, which results in a lower rate of climb.

For this reason, only a small takeoff flap setting is used for the takeoff to maintain an adequate airborne climb performance. However, large takeoff flap settings may be used to reduce the ground takeoff run as much as possible as long as the climb performance to the screen height is not compromised when the field length (runway) is limiting or the runway surface is poor. Conversely, no flaps may be used when takeoff distance is limiting.

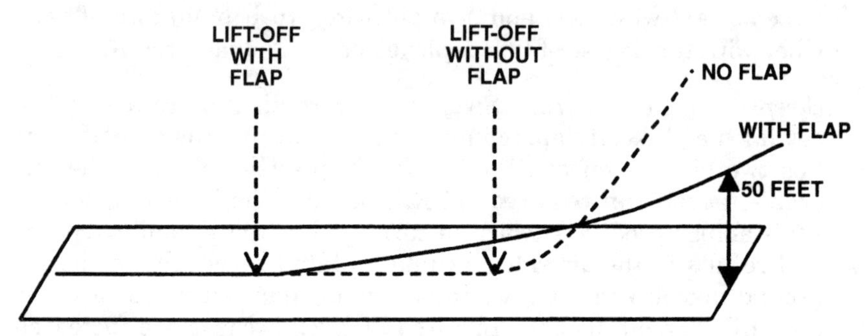

Figure 7.4 Effect of flaps on the takeoff ground run.

For large flap settings above the takeoff range, e.g., >20 degrees, the increased aerodynamic drag during the takeoff ground run causes a slower acceleration that results in an unacceptable increase in the takeoff run required and then once airborne significantly reduces the lift-drag ratio, which degrades the climb gradient performance to an unacceptable level.

The principles of flap deployment on the takeoff performance for both swept- and straight-wing aircraft is similar, but the effects are much more acute for swept-wing aircraft. Therefore, the use of takeoff flap deployment is much more rigid on swept-wing aircraft, whereas straight-wing (turboprop) aircraft have more flexibility and variation among different aircraft types.

How does pressure altitude affect takeoff performance?

A high aerodrome elevation (high pressure altitude) decreases an aircraft's performance and results in an increased takeoff distance required (TODR). Therefore, *high* means a decrease in performance that results in either more takeoff distance required or a lower takeoff weight.

How does air density (rho)/density altitude (pressure altitude and temperature) affect the takeoff performance?

An increase in density altitude (decrease in air density) increases the takeoff distance required (TODR). Therefore, *hot* and *high* mean a decrease in performance that results in either a greater TODR or a lower TOW.

How does humidity affect takeoff performance?

High humidity decreases air density, which decreases an aircraft's aerodynamic (C_L) and engine performance and results in an increased TOR/D required for a given aircraft weight. (*See Q: How does density altitude affect the takeoff performance? page 198.*) Therefore, *hot, high,* and *humid* mean a decrease in performance that results in either a greater TODR or a lower TOW.

How does wind affect the takeoff performance?

Wind has a profound effect on the takeoff performance of an aircraft. An aircraft may experience either a headwind, tailwind, or crosswind. (*See Q: What are the crosswind limitations on an aircraft? page 200.*)

Headwind. A headwind reduces the takeoff distance required for a given aircraft weight or permits a higher TOW for the TOR/D available. Thus the greater the headwind, the better is the aircraft performance.

Tailwind. A tailwind increases the takeoff distance required for a given aircraft weight or requires a lower TOW for the TOR/D available. Thus the greater the tailwind, the worse is the aircraft performance. It is for this reason that a takeoff with a tailwind shows very poor airmanship.

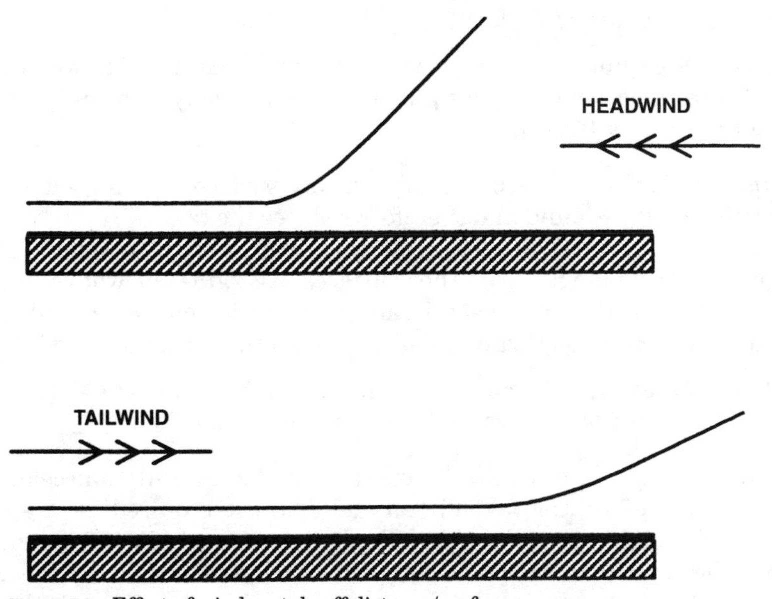

Figure 7.5 Effect of wind on takeoff distance/performance.

What are the recommended adjustments to headwind and tailwind components when calculating the takeoff and landing field length performance?

Not more than 50 percent of the reported headwind or not less than 150 percent of the reported tailwind should be used to calculate the takeoff or landing performance.

These adjustments provide a safety margin to the reported wind that covers acceptable fluctuations of the actual wind experienced.

What are the crosswind limitations on an aircraft?

The aircraft must not take off or land in a crosswind that exceeds its certified maximum crosswind limitation for the aircraft type to safeguard directly the directional and lateral control and indirectly the takeoff run performance of the aircraft.

For crosswind component calculations, see Chapter 9, "Flight Operations and Technique," page 283.

How does the runway length, surface, and slope affect takeoff performance?

Runway length. The length of the available runway is one of the performance limitations that restricts the maximum weight of the aircraft. The greater the runway length, the greater is the acceleration the aircraft can gain, and higher is the liftoff speed (V_R) the aircraft can obtain. And because the V_R speed is related to aircraft weight, it can be seen that the longer the runway available, the greater is the possible aircraft takeoff weight.

Note: Other limitations, such as the maximum permissible structural certificate of airworthiness weight of the aircraft may be more limiting than the runway length.

Runway surface. A hard and dry runway surface allows good acceleration on the ground and therefore reduces the takeoff run required for a given aircraft weight or allows a higher aircraft weight for a given runway length. On other surfaces, e.g., grass or wet contaminated hard surfaces, the acceleration is retarded on the ground, and therefore, the takeoff run and distance required are increased.

Note: Takeoff performance graphs and tables (i.e., field length) normally allow for such variations in surface conditions.

Runway slope. A downward slope allows the aircraft to accelerate faster; therefore, the takeoff run and distance required for a given aircraft weight are reduced or a higher maximum takeoff weight is possible for a given runway length compared with a level runway. An upward slope hinders the aircraft's acceleration; therefore, the takeoff run and distance required for a given aircraft weight are increased or the aircraft maximum takeoff weight is reduced for a given runway length compared with a level runway.

Note: Takeoff performance graphs and tables (i.e., field length) normally allow for such variations in runway slope.

When are you not permitted to take off from a wet runway?

You are not permitted to take off from a wet (contaminated) runway when

1. The aircraft antiskid system is inoperative.
2. The standing water level is above a specified limit.
3. Any other type-specific restrictions.

Describe field length limits.

The most restrictive field length available from either

1. The all-engine-operating runway length
2. The runway emergency distance length available, or
3. The one-engine-inoperative runway length

limits the aircraft's MTOW (or MLW) so that it meets the required TOR/D performance given the ambient aerodrome conditions of pressure altitude and temperature (density).

Describe the weight, altitude, and temperature (WAT) limits.

The weight, altitude, and temperature (WAT) conditions limit the aircraft's MTOW (or MLW) so that it meets the required second segment (and missed approach) climb gradient performance with one engine inoperative given the ambient aerodrome conditions of pressure altitude and temperature (density).

What guaranteed altitude/height would you be able to achieve at MTOW WAT-limited conditions with one engine inoperative?

The minimum height an aircraft would be able to achieve given these conditions would be the circuit height, i.e., 1500 ft.

What is an assumed/flexible temperature?

An assumed/flexible temperature is a performance calculation technique used to find the takeoff engine pressure ratio (EPR) setting for an aircraft's actual takeoff weight. This is known as a *reduced/derated thrust value.*

First, it should be clearly understood that the full takeoff thrust is calculated against an aircraft's performance limited (either field length, WAT, tire or net flight path, obstacle clearance climb profile) maximum permissible takeoff weight (MTOW), which itself is calculated from the ambient conditions of aerodrome pressure altitude and temperature. However, in many cases, the aircraft takes off with a weight lower than the maximum permissible takeoff weight. When this happens, an assumed temperature performance technique presents a method of calculating a decreased takeoff thrust that is adapted for the aircraft's actual takeoff weight. This is done by calculating the corresponding assumed/flexible temperature (higher than the actual air temperature) from the weight, altitude, and temperature (WAT) performance graph by using the aircraft's actual takeoff weight as if it were the performance-limiting MTOW against the actual aerodrome altitude to find the limiting temperature.

Note: Temperature increases as the maximum permissible weight decreases, so it is possible to assume a temperature at which the actual takeoff weight would be the limiting one.

Then, by using this calculated assumed/flexible temperature, you can calculate the aircraft's correct takeoff EPR thrust from the takeoff EPR graph. The assumed temperature (actual aircraft weight) EPR is a reduced/derated thrust value from the MTOW full-thrust setting. (*See Q: What is a variable/reduced-thrust takeoff? page 75.*)

Note: Assumed/flexible temperature setting for lower-weight aircraft meet the same TOR performance (i.e., rotate at the same point) as the maximum thrust setting for MTOW aircraft.

Assumed temperatures can be used to

1. Calculate on paper the required takeoff EPR
2. Enter into the aircraft's engine computer to calculate the required takeoff EPR

Finally, remember that the performance condition that was limiting at full thrust (field length, WAT curve, net flight path) is not necessarily the limiting condition at a reduced thrust.

What are the limitations of a variable/reduced-thrust (flexible) takeoff?

(*See Chapter 2, "Engines," page 77.*)

Describe tire speed limits.

Tire speed limit restricts the aircraft's maximum takeoff weight (MTOW) given the ambient aerodrome pressure altitude, temperature, and wind conditions so that the V_R (liftoff speed) is less than the maximum rated ground speed limit for the tires to protect against the tires blowing out during the takeoff roll.

Describe brake energy limits.

The brake energy capacity limits the aircraft's maximum takeoff weight (MTOW) given the ambient aerodrome pressure altitude, temperature, wind, and runway slope conditions so that V_1 does not exceed VMBE to ensure that the aircraft's brake system has sufficient energy to dissipate and stop the aircraft's inertia from V_1 under most operating conditions. (*See Q: What is VMBE speed? page 191.*)

How are reverse thrust, antiskid, and braking applied to stopping distance?

Reverse thrust. In general, the performance gained by using reverse thrust is not applied to takeoff emergency stopping distance (EMDR) or landing stopping distance, although a 10 percent safety factor is commonly applied to landing distance in the event of an inoperative thrust reverser.

Antiskid. The performance gained by using the antiskid system is applied to both takeoff distance and landing stopping distance. If the antiskid system is inoperative, then takeoff from a wet runway is normally prohibited, and the landing calculation has a large safety factor calculated to its landing distance required, usually about 50 percent.

Braking. Maximum braking performance is applied to both takeoff emergency stopping distance (EMDR) and landing distance required.

Describe the difference between net and gross flight paths/performance.

The difference between the net and gross flight paths/performance is as follows: The gross performance is the average performance that a fleet of aircraft should achieve if maintained satisfactorily and flown in accordance with the techniques established during flight certification and subsequently described in the aircraft performance manual. Gross performance therefore defines a level of performance that any aircraft of the same type has a 50 percent chance of exceeding at any time.

The net performance is the gross performance diminished to allow for various contingencies that cannot be accounted for operationally, e.g., variations in piloting technique, temporarily below-average performance, etc. It is improbable that the net performance/flight path will not be achieved in operation, provided the aircraft is flown in accordance with the recommended techniques, i.e., power, attitude, and speed.

Normally, performance graphs show net performance; however, some performance charts, especially for three-engined aircraft, assume gross is equal to net performance. This means that no margin exists between what the graph suggests the aircraft will achieve and what the aircraft will achieve.

Describe the departure profile segments (sectors) 1 to 4.

The various segments and other terms relating to the takeoff flight path are as follows:

Reference zero. Defined as the ground point at the end of the takeoff distance, below the net takeoff flight path screen height.

First (sector) segment. Extends from the reference point (35-ft height) to the point where the landing gear is retracted at a constant V_2 speed.

Second (sector) segment. Extends from the end of the first segment to a gross height of between at least 400 ft and a usual maximum of 1000 ft above ground level (AGL) at a constant V_2 speed.

Third (sector) segment. Assumes a level-flight acceleration during which the flaps are retracted in accordance with the recommended speed schedule.

Fourth and final (sector) segment. Extends from the third segment level-off height to a net height of 1500 ft or more with flaps up and maximum continuous thrust.

What is climb gradient?

Climb gradient is the ratio, in the same units, expressed as a percentage of change in height divided by horizontal distance traveled. The climb gradients on performance charts are true gradients for the all-up weight (AUW) of the aircraft, which allows for temperature, aerodrome pressure altitude, and aircraft configuration. That is, they are achieved from true rates of climb, not pressure.

Describe the net takeoff flight path (obstacle clearance).

The net takeoff flight path is the true height versus the horizontal distance traveled from reference zero, assuming failure of the critical engine at V_1, and it is used to determine the obstacle clearance by a specified minimal amount, normally 35 ft.

What is the typical jet takeoff technique and the various flight path options?

The takeoff flight path is based on the following technique and options: The flight path begins at reference zero, where the jet aircraft has attained a height of 35 ft (in dry conditions) and V_2 after failure of its critical engine at V_1. Landing gear retraction is completed at the end of the first segment, and the climb is continued at V_2 with takeoff flaps and takeoff thrust on the operating engine(s) until the second- to third-segment level-off transition takes place at either (1) a minimum gross height of 400 ft or (2) the maximum standard gross height, usually 1000 ft.

After option 1 or 2, the aircraft is then accelerated in level flight, flaps are retracted, and acceleration is continued to the final segment, where the climb is reassumed on maximum continuous thrust or (3) the maximum height, which can be reached using either (1) takeoff thrust for its maximum period after the brakes are released, usually 5 minutes, or (2) maximum continuous thrust to an unrestricted height

Option 3 is used to clear distant obstacles, i.e., in the normal third segment using an extended V_2 technique to gain a greater climb gradient. (*See Qs: What is an extended V_2 climb? page 207; Why is an extended V_2 climb used? page 207.*)

What does it mean if a takeoff weight is limited by an obstacle in the second segment?

If an aircraft is said to be takeoff-weight-limited by an obstacle in the second segment, this means that the aircraft's takeoff weight has to be reduced to ensure an adequate climb performance at a normal V_2 speed to clear any obstacles below the second- to third-segment level-off height.

What can you do if an aircraft is limited by a close-in obstacle in the second departure profile segment?

If an aircraft is limited by an obstacle in the second segment because its climb performance is insufficient to clear the obstacle(s), then either of the following flight procedures or a combination thereof can be used to improve the aircraft's climb gradient and thus clear the close-in obstacle(s):

1. Increase takeoff flaps (remaining in the takeoff flap range)
2. Reduce takeoff weight to a level that achieves the required climb gradient that clears the obstacles
3. Increased V_2 climb, maintaining takeoff weight (*See Q: What is an increased V_2? page 206.*)
4. Maximum angle climb profile [*See Q: Describe maximum angle (V_x) climb profile, page 207.*]

What is an increased V_2?

An increased V_2 is a technique for improving an aircraft's climb gradient performance in the second segment by increasing the V_2 climb speed. That is, increasing the V_2 base speed increases lift because lift is a function of speed, and for a given weight, an increased V_2 speed will provide an increase in lift, thus producing a greater net climb gradient. Thus an increased V_2 climb allows higher obstacles to be cleared in the second segment climb-out profile.

Note: An increased V_2 speed requires either a greater than normal takeoff distance (TOD) or a reduced aircraft weight.

When is an increased V_2 climb profile used?

An increased V_2 climb profile technique can only be used if the takeoff weight is not restricted by field length limits so that some or all of the excess field length can be used to increase takeoff speed V_R and thus V_2 speed.

Increased V_2 technique is used for the following reasons:

1. To achieve a greater obstacle clearance performance with an improved takeoff climb gradient without reducing takeoff weight
2. To allow a higher aircraft takeoff weight that achieves the standard (minimum) takeoff climb gradient corresponding to the required obstacle clearance gradient

The use of increased V_2 techniques usually is prohibited on wet runways.

What can you do if an aircraft is limited by a distant obstacle in the third departure profile segment?

If an obstacle in the third segment limits an aircraft because the obstacle in question is higher than the second- to third-segment level-off height, then either of the following flight procedures or a combination thereof can be used to clear the distant obstacle(s):

1. Extended V_2 climb profile technique (*See Q: What is an extended V_2 climb? page 207.*)
2. Reduce takeoff weight to a level that achieves the required climb profile and clears the obstacle
3. Flight path climbing turns to avoid the obstacles

Note: The distance and height of the obstacle dictate the procedures available. However, these vary with different aircraft types because of different performance capabilities.

What is an extended V_2 climb?

An extended V_2 climb is one in which the aircraft's second-segment climb; i.e., at V_2 and takeoff flaps, is either (1) continued to the highest possible level-off height, allowing for acceleration and flap retraction, if applicable, which can be reached with maximum takeoff thrust on all operating engines for its maximum time limit, normally 5 minutes, or (2) continued to an unlimited height with maximum continuous thrust, instead of takeoff thrust, that meets the aircraft's minimum acceleration and climb gradient requirements in the takeoff flight path above 400 ft.

Why is an extended V_2 climb used?

An extended V_2 climb is used to clear distant third-segment obstacles that are higher than the normal second- to third-segment level-off height, which is typically between 400 and 1000 ft.

An extended V_2 may only be used to clear the last obstacle in the flight path so that a normal third-segment acceleration and final-segment profile can be achieved. (*See Q: What are the typical jet takeoff technique and various flight path options? page 204.*)

Describe a maximum-angle (V_x) climb profile.

Maximum- or best-angle climb (V_x) is the steepest angle or highest gradient of climb used to clear close-in obstacles over the shortest horizontal distance.

Describe the minimum-rate (V_y) climb profile.

Maximum- or best-rate climb (V_y) is the highest vertical speed that gains height in the shortest time.

Describe the cruise climb profile.

The cruise climb profile is a compromise between the best en route speed profile and the best climb profile—most commonly used by commercial traffic. It provides faster en route performance, a more comfortable aircraft attitude, better aircraft control due to lower angle of attack, and greater airflow over the control surfaces.

What climb departure uses the least trip fuel?

The best rate of climb (V_y) departure uses the least trip fuel because it ensures that the aircraft reaches its optimal cruise altitude as quickly as possible, and therefore, the aircraft spends a greater part of its flight time at its optimal altitude than with any other climb profile.

The optimal en route altitude has the best aerodynamic and engine performance qualities. By being at this optimal cruise altitude as long as possible, the best fuel economy and specific fuel consumption (SFC) are obtained for the largest percentage of the flight (trip), and therefore, the least trip fuel is used. (*See Q: Why does a jet aircraft climb as high as possible? page 69.*)

A reduced-power climb uses more or less trip fuel, and why?

A reduced-power climb uses more trip fuel. This is so because using a reduced-power climb means that the aircraft has a slower rate of ascent and therefore takes longer to reach its cruise altitude. Consequently, it spends less time at its optimal cruise altitude and thus uses more trip fuel.

Note: Remember that at its optimal cruise altitude an aircraft experiences the best aerodynamic and engine performance, which results in the best fuel economy (SFC). (*See Q: Why does a jet aircraft climb as high as possible? page 69.*)

A derated takeoff will use more or less trip fuel, and why.

A derated takeoff uses more trip fuel. This is so because, as with a reduced-power climb, the aircraft has a slower initial rate of ascent and therefore takes longer to reach its transition to its en route climb profile and then its cruise altitude. Consequently, it spends less time at its optimal cruise altitude and therefore uses more trip fuel.

Note: Remember that at its optimal cruise altitude an aircraft experiences the best aerodynamic and engine performance, which results in the best fuel economy (SFC). (*See Q: Why does a jet aircraft climb as high as possible? page 69.*)

En Route Performance and Flight Planning

What is VRA/MRA speed?

VRA/MRA is an airspeed for rough air conditions, or turbulence penetration speed. The rough airspeed is recommended for flight in turbulence that is based on the aircraft's V_B speed (design speed for maximum gust intensity). It provides speed protection against the two possibilities that stem from the effects of a disturbance in rough air conditions. In other words, the VRA/MRA speed is high enough to allow an adequate margin between the aircraft stall speed and also low enough to protect against structural damage from a high-speed gust disturbance.

What is VMO/MMO speed?

VMO/MMO (IAS velocity/Mach) is the maximum operating speed permitted for all operations. It is normally associated with jet aircraft.

What is VNO speed?

VNO (IAS velocity) is the normal operating speed permitted for all normal operations. It is normally associated with propeller aircraft.

What is VDF/MDF speed?

VDF/MDF (IAS velocity/Mach) is the maximum flight diving speed for a jet aircraft. It is established as the highest demonstrated speed during flight certification trials.

What is VNE speed?

VNE (IAS velocity) is the never-exceed velocity. It is associated with propeller-driven aircraft and is a higher speed than the VNO speed, which can be used when operationally desired but must never be exceeded.

What is the absolute ceiling?

The absolute ceiling is an aircraft's maximum attainable altitude/flight level at which the Mach number buffet and prestall buffet occur coincidentally. This scenario is known as *coffin corner.* (*See Q: What is coffin corner? page 26.*) Therefore, an aircraft is unable to climb above its absolute ceiling. The absolute ceiling is determined during flight certification trials.

What is a maximum service ceiling?

The maximum service ceiling is an aircraft's' imposed en route maximum operating altitude/flight level, which provides a safety margin below its absolute ceiling.

Define maximum endurance and range with reference to the drag curve.

The total drag generated by an aircraft is high at both high and low airspeeds. At high airspeeds the total drag is high because the aircraft experiences a lot of profile drag, and at low airspeeds the total drag is high because the aircraft experiences a lot of induced drag. Minimum drag occurs at an intermediate speed (VIMD). This is presented by an aircraft's drag curve. (*See Qs: What is drag? page 6; Describe the two major types of drag and their speed relationship, page 6; Describe the drag curve on a propeller/jet aircraft, pages 7 and 8.*)

Maximum endurance. This is achieved by flying at the maximum-endurance airspeed, which is the indicated airspeed that relates to the thrust required to balance the minimum drag (VIMD) experienced by the aircraft. Minimum drag (VIMD) is the lowest point/airspeed on the drag curve. And given that thrust is a product of engine power and fuel consumption is a function of the engine power used, the aircraft thus has the lowest fuel consumption in terms of pounds of fuel used per hour and hence produces the longest flight time for a given quantity of fuel when flying at its maximum endurance airspeed. Broadly speaking, this speed remains constant with altitude and is used for any delaying action, e.g., holding at a destination airfield.

Maximum range. This is achieved by flying at the maximum-range airspeed. This is the indicated airspeed that balances a value of drag slightly higher than the minimum drag point (best endurance speed) that achieves a greater range for a given quantity of fuel because the benefits of the increased airspeed outweighs the associated increase in drag and fuel consumption.

The maximum-range airspeed is where the ratio of power required (to produce thrust to balance drag) to airspeed is least.

Note: Airspeed is the rate of covering distance, and power required is the rate of burning fuel to produce thrust to balance drag.

This ratio occurs on the total drag curve where the line from the origin is a tangent to the curve. It is more common to use the maximum-range speed en route.

What is the difference between maximum-range cruise (MRC) and long-range-cruise (LRC)?

Maximum-range cruise. This is the speed at which, for a given weight and altitude, the maximum fuel mileage is obtained. It is difficult to establish and maintain stable cruise conditions at maximum-range speeds.

Long-range cruise. This is a speed significantly higher than the maximum-range speed, i.e., 10 knots (M 0.01), which results in a 1 percent mileage loss at a constant altitude. The long-range cruise schedule requires a gradual reduction in cruise speed as gross weight decreases with fuel burnoff.

What is a cost index?

A cost index (CI) is a performance management function that optimizes the aircraft's speed for the minimum cost.

Cost indices form part of a company's stored route and are inserted into the flight management computer (FMC). They take into account specific route factors such as the price of fuel at the departure and destination airports so that the aircraft is flown at the correct speed to balance the fuel costs against the dry operating costs. An incorrect CI will always cost more money.

How is range increased when flying into a headwind?

Range is increased when flying into a headwind because the best range speed will be a little faster, and the airspeed represents the rate of distance covered. Therefore, range will be increased with a headwind. The increased fuel flow is compensated for by a higher speed, allowing less time en route for the headwind to act.

A flight carried out below its optimal altitude has what results on jet performance?

A flight carried out below its optimal altitude uses more fuel but takes less time to complete the trip when flying at a *constant Mach number* (MN).

Note: Optimal altitude is the aircraft's highest attainable altitude, i.e., service ceiling, which for a jet typically would be an altitude above flight level 260, and thus the MN speed flown is constant because the MN is the limiting speed.

Note: Higher FL means that time increases and fuel consumption decreases (in still air and at a constant MN). Lower FL means that time decreases and fuel consumption increases (in still air and at a constant MN).

The aircraft uses more fuel because the engines are designed to achieve their best specific fuel consumption (SFC) at a high operating rpm, which can only be achieved at high (optimal) altitudes. (*See Q: Why does a jet aircraft climb as high as possible? page 69.*)

Therefore, when a flight is carried out below its optimal altitude, its engines cannot operate at their optimal rpms, and therefore, the SFC is higher. Consequently, the aircraft uses more fuel to complete the trip.

It takes less time because by flying at a constant MN, which is normal practice at high altitudes, i.e., above FL260, the local speed of sound (LSS) increases the lower the altitude because LSS is a function of temperature. [*See Qs: Describe the local speed of sound (LSS), page 122; How does temperature affect local speed of sound (LSS)? page 122.*]

Therefore, to maintain the same MN speed the indicated airspeed (IAS) and true airspeed (TAS) must be increased, and thus the ground speed increases (assuming a constant wind velocity at both altitudes). This results in the distance covered taking less time when a flight is carried out below the optimal altitude. For example,

$$\text{FL 370/MN 0.83} = \frac{\text{TAS 500}}{\text{LSS 600}}$$

$$\text{FL 350/MN 0.83} = \frac{\text{TAS 514}}{\text{LSS 620}}$$

Note: Flights at lower altitudes, i.e., below FL260, *where IAS is the reference speed,* will have the opposite effect. This is so because the TAS decreases at lower altitudes for a constant IAS; therefore, ground speed is reduced, and the trip will take a longer time to complete. However, a jet aircraft normally would fly at an altitude where the MN is the reference speed.

What is a cruise (step) climb?

A cruise (step) climb occurs when an aircraft in the cruise loses weight due to fuel burn, which allows the aircraft to fly higher; therefore, a cruise (step) climb is initiated to climb the aircraft to its new maximum altitude.

Note: An aircraft may have several step climbs, especially on long-haul routes.

A cruise climb is important because a jet aircraft's most efficient performance is gained at its highest possible altitude.

What are the normal en route operating performance limitations for an aircraft?

The normal en route operating performance limitations for an aircraft are

1. En route obstacle/terrain clearance with one or two engines inoperative
2. Maximum range limit
3. Extended twin operations (ETOPS) time limit

Explain a typical fuel plan for a trip.

A typical fuel profile for a flight would include sufficient fuel for the following:

1. Takeoff and climb at takeoff thrust
2. Climb to the initial cruise altitude at maximum continuous thrust

3. En route cruise, including intermediate step climbs
4. Descent to a diversion point (go-around point) over the destination aerodrome
5. Contingency fuel
6. Diversion to over an alternative aerodrome
7. An instrument approach and landing
8. An additional amount of holding fuel may be required if the destination is an island with no alternative or you are arriving at a major airport at its busiest period, e.g., London Heathrow between 0700 and 0930.

What is island holding fuel, and when is it used?

Island holding fuel is a quantity of fuel uplifted usually in place of diversion fuel that allows an aircraft to hold over a destination aerodrome for an extended period of time. It is often associated with sorties to remote islands, e.g., Easter Island, where there is no diversion option and there is a possibility of a delayed landing due to adverse weather patterns.

Explain fuel howgozit?

Fuel howgozit is a comparison of the actual fuel remaining against the planned fuel remaining along the flight path. At any stage of flight an estimate of the fuel difference can be obtained, and it is ideal to show any trends, such as increasing fuel burn.

Note: Fuel howgozit charts also can be used to plan the position of the point-of-no-return (PNR) point.

What is the critical point?

The critical point, or equal-time point, is the en route track position where it is as quick (time) to go to your destination as it is to turn back.

The critical point (CP) is calculated as a distance and time from the departure airfield using the following formula:

$$\text{Distance to CP} = \frac{DH}{O + H}$$

where D is total sector distance, H is ground speed home, and O is ground speed out.

$$\text{Time to CP} = \text{distance to CP}/O$$

Note: A critical point for an engine-out scenario is different from the all-engine operating critical point and usually is used to determine a worst-case scenario.

How does the wind affect the position of the critical point?

The critical point moves into the wind.

Given still air conditions, the critical point (CP) between two aerodromes is simply the halfway point. However, the effect of wind displaces the critical point to one side or the other of the midpoint. Flying into a headwind moves the CP closer to the destination aerodrome, and flying with a tailwind moves the CP closer to the departure aerodrome. This is so because the CP is the equal time point to reach an airfield, and therefore, ground speed is all-important, and ground speed is TAS × WV.

What is the most important diversion question to ask in an emergency?

The most important question to ask in an emergency given two diversion aerodromes is, Which aerodrome is the quickest to get to?

What is the point of no return (PNR)?

PNR (point of no return) is the last point on a route at which it is possible to return to the departure aerodrome with a sensible fuel reserve.

Normal PNR points are based on the aircraft's safe endurance. The all-engine PNR formula is

$$\text{Time to PNR} = \frac{EH}{O \times H}$$

where E is the safe endurance time, H is the ground speed home, and O is the ground speed out.

$$\text{Distance to PNR} = \text{time to PNR} \times \text{O}$$

The one-engine PNR distance is calculated as follows:

1. Calculate a 1-nautical-mile round-trip fuel flow (all-engine ground speed out and one-engine-inoperative ground speed home)
2. Safe endurance divided by 1-nautical-mile round trip fuel flow

One-engine PNR time is calculated by distance divided by ground speed out.

Where are you likely to need a point of no return (PNR)?

PNR calculations are important for aircraft for which diversion airfields are not readily available, e.g., over large water areas such as the Pacific Ocean. It is crucially important to have a PNR point if you elect to carry island holding fuel instead of diversion fuel because you become solely committed to landing at your destination once you pass the PNR point.

Note: PNR points are not really required on a route over land with diversion airfields available en route.

Descent and Landing

What is the landing distance available (LDA)?

LDA is the distance available for landing, taking into account any obstacles in the flight path, from 50 ft above the surface of the runway threshold height (fence) to the end of the landing runway.

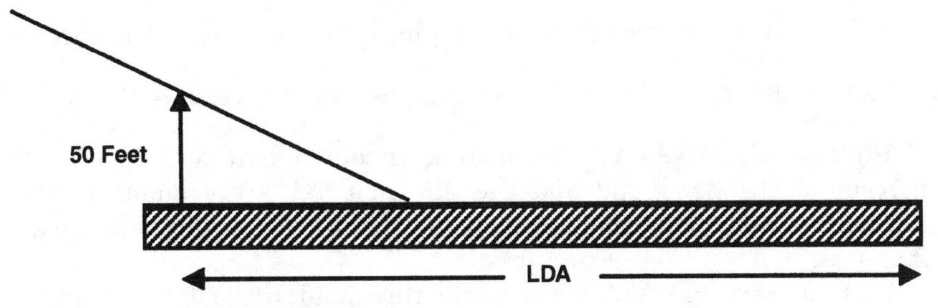

Figure 7.6 Landing distance available.

What is landing distance required (LDR)?

Landing distance required (LDR) is the distance required from the point where the aircraft is 50 ft over the runway (i.e., threshold fence height at a maximum VAT threshold speed) to the point where the aircraft reaches a full stop.

What distance along the runway is the touchdown aiming point?

1000 ft, given a 3-degree glide path. This is normally distinguished by marker board markings on the runway surface.

What is the height of the aircraft over the runway threshold (fence)?

Given a 3-degree glide path and a 1000-ft touchdown, the height of the aircraft over the runway threshold fence is 50 ft. However, larger aircraft will have a reduced clearance height due to their extended wheels.

What is VMCL speed?

VMCL is the minimum control speed in the air for a multiengined aircraft in the approach and landing configuration, at and above which it is possible to maintain directional control of the aircraft around the normal/vertical axis by use of the rudder after failure of an off-center engine within defined limits while applying variations of power, i.e., idle to maximum thrust on the live engine(s).

What is VAT/Vref speed?

Velocity at threshold (VAT)/velocity reference speed (V_{ref}) is the target approach threshold speed above the fence height for a specified flap setting that ensures that the landing field length is constantly achieved.

$VAT/V_{ref} = 1.3\ V_S$ in the landing configuration

VAT_0 = the target threshold, all-engine operation, speed with maximum flap setting

VAT_1 = the target threshold, one engine (critical) inoperative, speed

VAT_2 = the target threshold, two engines inoperative, speed

Adjustments to VAT/V_{ref} are made for headwind values. That is, 50 percent of the headwind and the full gust value are added to the VAT/V_{ref} speed up to a maximum permitted predetermined limit, e.g., 20 knots.

Above a maximum VAT speed at the threshold, the risk of exceeding the landing field length is unacceptably high, and a go-around should be initiated.

How does a fast approach speed affect landing distance?

The landing performance of an aircraft is based on specified approach speeds (VAT/V_{ref}). If you approach for a landing at a speed higher than that specified, the landing distance probably will exceed that calculated. This can be very important if the landing distance available (LDA) is limiting.

Note: Actual approach speed below the specified approach speed should not be flown for safety reasons.

How does aircraft weight affect landing performance?

Increased aircraft weight results in a greater landing distance required (LDR).

An increased aircraft weight has the following variable effects that increase the landing distance required:

1. The stalling speed is increased with the higher aircraft weight, so the minimum approach speed, V_{ref}/VAT ($1.3\ V_S$), must be higher. A

higher approach speed requires a greater landing distance to come to a full stop.

2. The greater aircraft weight and higher approach speed result in a greater momentum of the landing aircraft and therefore require more distance to stop.
3. The increased weight means that the kinetic energy (½ m V_2) is higher, and the brakes have to absorb this greater energy, which increases the landing distance required.

Overall, an increase in weight increases the LDR. As a guideline, a 10 percent increase in weight requires a 10 percent increase in landing distance, i.e., a factor of 1:1.

How does the use of flaps affect an aircraft's landing performance?

Increased flap settings decrease the landing distance required.

The use of increased flap settings has the following variable effects that decreases the landing distance required (LDR):

1. Flap deployment increases lift and reduces the stalling speed (V_S). Therefore, the approach speed, VAT/V_{ref} (1.3V_S), is less, which results in a shorter landing distance required.
2. The higher the flap deployment, the greater is the aerodynamic drag that helps to slow the aircraft down, and this results in a shorter landing distance required.
3. The higher the flap setting and the steeper the approach path, the lower is the forward velocity and momentum on landing, which results in a shorter landing distance required.

How does pressure altitude affect landing performance?

A high aerodrome elevation (high pressure altitude) decreases an aircraft's performance, which results in an increased landing distance required (LDR). Thus *hot* means a decrease in performance that results in either a greater landing distance required (LDR) or a lower landing weight (LW).

How does air density (rho)/density altitude (pressure altitude and temperature) affect the landing performance?

An increase in density altitude (decrease in air density) decreases an aircraft's performance, which results in an increased landing distance required (LDR). Thus, *hot* and *high* mean a decrease in performance that results in either a greater LDR or a lower landing weight (LW).

How does humidity affect landing performance?

High humidity decreases air density, which decreases an aircraft's aerodynamic (C_L) and engine performance, resulting in an increased landing distance required (LDR) for a given aircraft weight. (*See Q: How does density altitude affect landing performance? page 217.*) Thus *hot, high,* and *humid* mean a decrease in performance that results in either a greater LDR or a lower landing weight (LW).

How does wind affect landing performance?

Wind has a profound effect on the landing performance of an aircraft. An aircraft may experience either a headwind, tailwind, or crosswind. (*See Q: What are the crosswind limitations on an aircraft? page 200.*)

Headwind. A headwind reduces the landing distance required (LDR) for a given aircraft weight, or it permits a higher landing weight (LW) for the landing distance available (LDA). This is so because the ground speed (GS) is reduced by the headwind (HW) for the same true airspeed (TAS). That is,

$$\text{TAS} = 120 \text{ knots}$$

$$\text{HW} = 20 \text{ knots}$$

$$\text{GS} = 100 \text{ knot}$$

Therefore, the aircraft has a lower touchdown speed from which it has to stop. Additionally, a headwind provides a resistance to the motion of the aircraft that is used as an efficient braking drag. These effects result in a shorter LDR for a given aircraft weight. Thus the greater the headwind, the greater is the aircraft's landing performance.

Tailwind. A tailwind increases the LDR for a given aircraft weight, or it requires a lower LW for the LDA. This is so because the ground speed is increased by the tailwind (TW) for the same TAS. That is,

$$\text{TAS} = 120 \text{ knots}$$

$$\text{HW} = 20 \text{ knots}$$

$$\text{GS} = 140 \text{ knots}$$

Therefore, the aircraft has a higher touchdown speed from which it has to stop. Additionally, the braking drag is less effective; in fact, a tailwind acts as a push effect on the aircraft, which is completely the opposite effect required to assist a landing aircraft. Therefore, a longer landing distance is required. Thus the greater the tailwind, the worse is the aircraft's landing performance.

What are the recommended adjustments to headwind and tailwind components when calculating the takeoff and landing field length performance?

(*See page 200.*)

What are the crosswind limitations on an aircraft?

(*See page 200.*)

How does the runway length, surface, and slope affect the landing performance?

Runway length. The length of the available runway is one of the performance limitations that restrict the maximum weight of the aircraft. The longer the runway length and the greater the aircraft's stopping action, the higher its VAT/V_{ref} approach speed can be. And because the VAT/V_{ref} speed is related to the aircraft's weight, it can be seen that the longer the runway length available, the greater the aircraft's landing weight can be.

Note: Other limitations, such as the maximum permissible structural certificate of airworthiness weight of the aircraft may be more limiting than the runway length.

Runway surface. Low-friction surfaces increase landing distance required (LDR) for a given aircraft weight because the surface does not permit effective braking.

Note: A low-friction surface can be a contaminated hard surface; e.g., snow and standing water, or a nonhard surface, e.g., grass and mud surfaces.

A particular runway surface phenomenon that significantly increases landing distance (LD) is aquaplaning. (*See Q: What is aquaplaning? page 321.*) Because of the possibility of aquaplaning on a wet runway, landing into a headwind is highly recommended.

Note: Landing performance graphs and tables (i.e., field lengths) normally allow for such variations in surface conditions.

Runway slope. A downward slope requires a longer LD for a given aircraft weight or a lower landing weight (LW) for a given landing distance available (LDA). This is so because (1) it takes longer for the aircraft to touch down from the 50-ft threshold (fence) height because the runway is falling away beneath the aircraft, and (2) the downslope maintains the aircraft's momentum, and therefore, the aircraft's braking is less effective. An upward slope requires a shorter LD for a given aircraft weight or allows a higher LW for a given

LDA. This is so because (1) it takes a shorter distance for the aircraft to touch down from the 50-ft threshold (fence) height because the runway is rising beneath the aircraft and (2) the upslope dissipates the aircraft's momentum, and therefore, the aircraft's braking is more effective.

Note: Landing performance graphs and tables (i.e., field lengths) normally allow for such variations in runway slope.

What is RLW?

Restricted landing weight, i.e., the maximum landing weight for the runway length (LDA) and conditions.

What factors are taken into account on restricted landing weight (RLW)?

Engine-out overshoot performance, weight, altitude, and pressure (WAT), runway length (LDA) and conditions.

Chapter

8

Meteorology and Weather Recognition

For atmosphere, air pressure, and density questions, see Chapter 5, "Atmosphere and Speed," page 115.

What is an isobar?

An isobar is a line on a meteorologic chart that joins places of equal sea level pressure.

What is an isotherm?

An isotherm is a line joining places of the same mean temperature.

Heat and Temperature

What is heat?

Heat is a form of energy that is measured in calories. One calorie represents the energy required to raise the temperature of 1 g of water 1°C, so the energy required to raise the temperature of 2 g of water by 1°C would be 2 calories.

What provides the earth's heat energy?

The sun in the form of solar radiation provides the earth's heat and light.

The solar radiation's wavelengths are such that only about half the radiation that hits the upper atmosphere finally reaches the earth's surface (the rest is either reflected back into space or absorbed into the upper layers of the atmosphere). This short-wave radiation is absorbed by the earth's surface, causing its temperature to increase. This is called *insulation.* Energy is then reradiated out from the ground as long-wave radiation (in the 10-μm band), and it is this radiation that heats up the lower atmosphere, near the ground, with the result that any parcel of air that is warmer than the surrounding air will rise. (*See Q: What are the different ways of transferring heat energy into the atmosphere? page 222.*)

What are the different ways of transferring heat energy into the atmosphere?

The heat held in the earth's surface can be transferred to the atmosphere by

1. Radiation
2. Conduction
3. Convection

These contribute considerably to the complexity of weather patterns.

How does cloud cover affect the heating of the earth's surface?

By day. Cloud cover stops the incoming solar radiation from penetrating to the earth's surface by reflecting it back into space, causing the heating of the earth's surface and thereby the lower layers of the atmosphere next to the ground to be reduced. This results in lower temperatures and less atmospheric convective movement.

By night. Cloud cover causes the opposite effect by trapping the heat energy in the lower atmosphere on and near the ground and reflecting it back to the surface, in effect, recirculating the heat. This results in the lower atmosphere, beneath the clouds, maintaining a higher temperature because it experiences less cooling than on a clear night.

What is specific heat capacity?

Specific heat capacity is the ability of a material to hold thermal/heat energy. (*See Q: What is heat? page 221.*)

What is latent heat?

Latent heat is the heat energy, measured in calories, absorbed or released when water changes from one state to another.

Note: There are three states of water:

1. Water vapor (gas)
2. Water liquid (cloud, mist, fog, rain, etc.)
3. Water solids (ice)

When water changes to a higher energy state, i.e., from ice to liquid to vapor, it absorbs/uses latent heat energy (from the surrounding atmosphere/properties). And when it moves to a lower energy state, i.e., from vapor to liquid to ice, it releases latent heat energy (into the surrounding atmosphere/properties). This can be important in certain forms of airframe ice formation. For example, 80 calories of latent heat is released with the change of state of 1 g of water to ice. If this water were supercooled to say −20°C, then it would already have 20 calories less latent heat energy than it ought to have at 0°C. When it then reverts to ice from water, this is made up by the latent heat being released by the change of state of the water to ice. Therefore, the 20 calories needed will change 20/80, or ¼, of the water to ice. The remaining ¾ of the water runs back along the airframe, rapidly cooling and freezing as a clear sheet of ice. (*See Q: What are supercooled water droplets? page 262.*)

What is temperature?

Temperature is a measure of molecule agitation in a substance, which is represented as the hotness of a body. Therefore, temperature can be thought of as a measure of the hotness of a body.

What factors determine the temperature at the earth's surface (i.e., why is it hot in the tropics and cold at the poles)?

The temperature at the earth's surface and the lower atmosphere where most of the weather is found depends on two factors:

1. How much heat energy in the form of short-wave energy (solar radiation) reaches the earth's surface. This depends on the following factors:
   a. *Latitude.* The more directly the sun's rays hit the earth's surface, the greater is the quantity of heat energy transferred and therefore the greater is its temperature. The sun is directly overhead in the tropics, and this is where the intensity of insulation is greatest.
   b. *Season.* The earth's axis of rotation is tilted with reference to its orbit. This tilting creates the same effect as a change of latitude and is the primary cause of the different seasons and associated temperature changes throughout the year.

   c. *Time.* The time of day determines the amount of the sun's heating on the earth's surface. The highest temperature of the day is not experienced until later in the afternoon, at approximately 1500 hours during the summer months.
2. The energy absorption (and retention) capacity of the surface.
   a. *Absorption.* The type of surface determines how much heat energy is absorbed, i.e., its reflective quality.
   b. *Specific heat capacity of the surface.* (*See Q: What is specific heat capacity? page 222.*)

What is the difference between Celsius and Fahrenheit?

Celsius and Fahrenheit are both scales used for measuring temperature. The Celsius scale (°C) divides the temperature between the boiling point and the freezing point of water into 100 degrees. That is, the boiling point of water is 100°C, and the freezing point of water is 0°C.

Note: The Celsius scale is the most commonly used form of temperature measurement. The Fahrenheit scale (°F), is based on water boiling at 212°F and freezing at 32°F.

What is the formula to convert Celsius and Fahrenheit?

The formula to convert Celsius to Fahrenheit is

$$°F = \frac{9}{5}(x°C + 32)$$

The formula to convert Fahrenheit to Celsius is

$$°C = \frac{5}{9}(x°F - 32)$$

Describe OAT.

OAT is the ambient outside air temperature.

Describe SAT.

SAT is the ambient static air temperature. This is commonly used as a different name for outside air temperature (OAT).

Describe TAT.

TAT is the total air temperature indicated on the air temperature instrument; it is a product of the static air temperature (SAT) and the adiabatic compression (ram) rise in temperature experienced on the temperature probe.

Note: Therefore, TAT is a higher temperature than outside air temperature (OAT) whenever there is an airflow into the temperature probe, which is sometimes referred to as a *heating error* when you need to calculate the actual OAT. This heating error can remain after a flight due to the residual heat left in the probe, which is why you sometimes have a difference in OAT and TAT shortly after a flight.

How does a change in air temperature affect an aircraft's flight level?

An air temperature that differs from the international standard atmosphere (ISA) temperature will result in a different actual flight level (i.e., height above the ground) than the pressure level read by the altimeter.

A higher than ISA air temperature makes the air less dense and lighter in weight, causing the density altitude to differ from the pressure altitude. This results in the actual flight level being higher than the pressure level read by the altimeter.

However, a lower than ISA air temperature makes the air denser and heavier in weight, causing the density altitude to differ from the pressure altitude. This results in the actual flight level being lower than the pressure level read by the altimeter. (*See Q: What density errors are commonly experienced? page 119.*)

Therefore, when flying from a high to low (temperature), beware below, because your actual flight level (and therefore ground clearance) is lower than indicated by your altimeter. In other words, your altimeter overreads.

This high-to-low mnemonic applies equally to pressure values as it does to temperature. (*See Q: How does a change in pressure affect the aircraft's flight level? page 249.*)

What is a temperature inversion/layer?

A temperature inversion occurs when the air closest to the ground, or even the ground itself, is cooler than the air above it. In other words, the air temperature increases with height (rather than the usual decrease).

A temperature inversion layer is the height/altitude where the air temperature changes from the temperature increasing with height (i.e., a temperature inversion) to a normal decrease in temperature with height state.

When a temperature inversion occurs, it acts like a blanket, stopping vertical movement/currents; i.e., air that starts to rise meets an inversion layer and so stops rising.

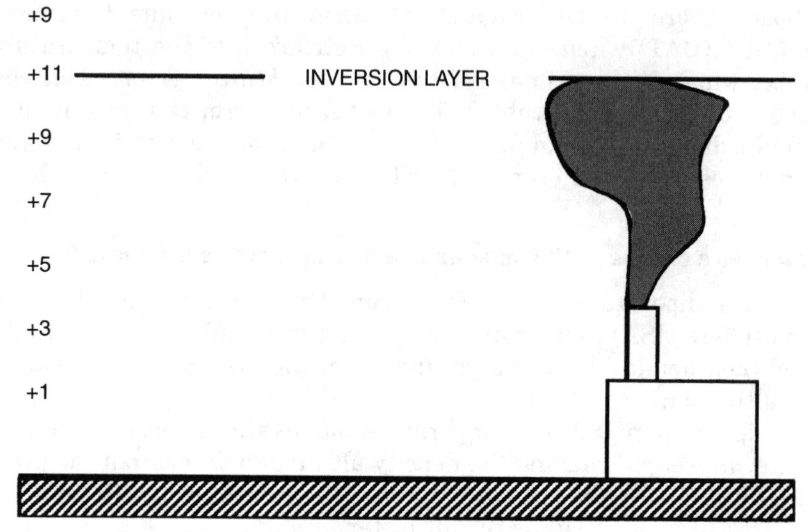

Figure 8.1 Temperature/inversion layer.

What is an isothermal layer?

An isothermal layer is one where the air remains at the same temperature through a vertical section of the atmosphere (i.e., a constant 10°C between 5000 and 10,000 ft).

Note: Remember that temperature usually decreases with altitude.

Moisture and Clouds

What is the adiabatic process?

The adiabatic process is one in which heat is neither added nor removed from a system, but any expansion or compression of its gases changes the temperature of the system with no overall loss or gain of energy. That is, compressing air increases its temperature, and decompressing it (expansion) reduces its temperature.

A common adiabatic process for pilots is the expansion of a cooling parcel of air when it rises in the atmosphere. (*See Qs: What is ELR/DALR/SALR? page 226 and 227; Explain relative humidity, page 227; What is dew point? page 227.*)

What is ELR?

Environmental lapse rate (ELR) is the rate of temperature change with height of the general surrounding atmosphere.

The international standard atmosphere (ISA) assumes an ELR of 2°C per 1000 ft of height/altitude gained. The actual ELR in a real atmosphere, however, may differ greatly from this; in fact, it can be zero (isothermal layer) or even a negative value (inversion).

What is DALR?

Dry adiabatic lapse rate (DALR) is the adiabatic temperature change for unsaturated air as it rises. Unsaturated air is known as dry air, and its change in temperature is a rather regular drop of 3°C per 1000 ft of height/altitude gained.

What is SALR?

Saturated adiabatic lapse rate (SALR) is the adiabatic change in temperature for saturated air as it rises.

SALR commences at a height where a parcel of air's temperature is reduced to its dew point temperature, 100 percent relative humidity, and its water starts to condense out to form a cloud. (*See Qs: Explain humidity/relative humidity, page 227; What is dewpoint? page 227.*)

Above this height, the now saturated air will continue to cool as it rises, but because it releases latent heat as the water vapor condenses into a liquid form, i.e., clouds, its cooling rate is reduced to a rather regular drop of 1.5°C per 1000 ft of height/altitude gained.

Explain humidity/relative humidity.

Humidity is the water vapor in the air, and relative humidity is a measure of the amount of water vapor present in a parcel of air compared with the maximum amount it can support (i.e., when the air is saturated) at the same temperature. Relative humidity is usually expressed as a percentage, i.e., relative humidity is 100 percent when the air is saturated.

How does air temperature affect relative humidity?

The amount of water vapor a parcel of air can hold depends on its temperature. That is, warm air is able to hold more water (in a vapor or liquid state) than colder air. In other words, cooler air supports less water vapor.

What is dewpoint?

The dewpoint is the temperature at which a parcel of air becomes saturated. That is, its capacity to hold water vapor is equal to that which it is actually holding, or in other words, its relative humidity is 100 percent.

Note: Dewpoint is also sometimes called the *saturated temperature.* The higher the moisture content in the air, the higher is its dewpoint temperature. (*See Q: Describe how clouds are formed, page 228.*)

Describe how clouds are formed.

For cloud formation to be possible, the following properties must exist:

1. Moisture present in the air.
2. A lifting action to cause a parcel of air to rise. The four main lifting actions are
   a. Convection
   b. Turbulence
   c. Frontal
   d. Orographic
3. Adiabatic cooling of the rising air.

If a parcel of air containing water vapor is lifted sufficiently, it will cool adiabatically, and its capacity to hold water vapor will decrease (i.e., cooler air supports less water). Therefore, its relative humidity increases until the parcel of air cools to its dewpoint temperature, where its capacity to hold water vapor is equal to that which it is actually holding, and the parcel of air is said to be *saturated* (i.e., its relative humidity is 100 percent). Any further cooling will cause some of the water vapor to condense out of its vapor state as water droplets and form clouds. Further, if the air is unable to support these water droplets, then they will fall as precipitation in the form of rain, hail, or snow.

How is the height of a cloud base determined?

Provided that the properties for a cloud to form are present (i.e., moisture, lifting action, adiabatic cooling, etc.), the height of a cloud base is determined by the difference between the dewpoint temperature and the ground temperature. The difference divided by the appropriate lapse rate per 1000 ft will determine the height of the cloud base. (*See Qs: Describe how clouds are formed, page 228; What is dewpoint? page 227; What is ELR/DALR/SALR? page 226 and 227.*)

Note: The higher the moisture content present in the rising air, the higher is the dewpoint temperature, and the less the difference between the surface and the dewpoint temperatures, the lower is the cloud base. Thus the amount of moisture content in the air is a determining factor of the cloud base height.

$$\text{Cloud base height (in ft)} = \frac{19^\circ\text{C (surface temp)} - 13^\circ\text{C (dewpoint)}}{3\text{(DALR)}} \times 1000$$

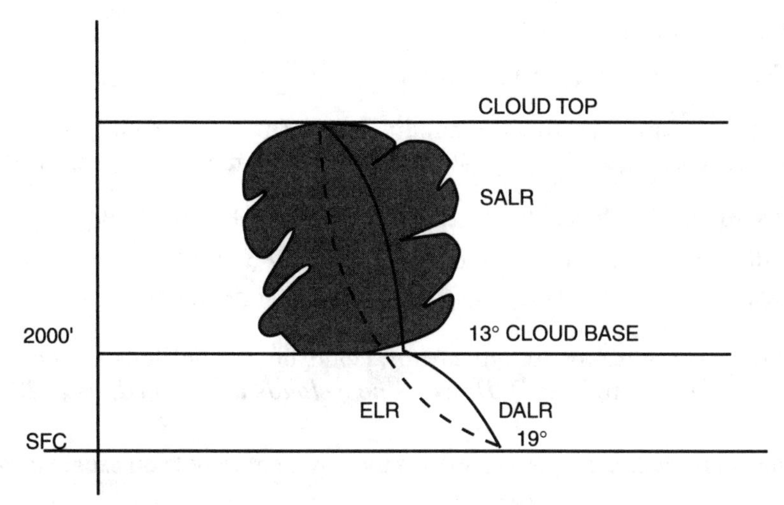

Figure 8.2 Cloud base.

It is worth noting that the relationship of the temperatures in the cloud and the surrounding air (i.e., the ELR/DALR/SALR) determines not only the type of cloud formation but also the height of the cloud top. That is, when the surrounding temperature is the same as or warmer than the temperature inside the cloud, then the air in the cloud will become stable and stop rising. This limits the height of the cloud top.

Why do cumulus clouds have flat bases and round tops?

Cumulus clouds have flat bases due to the uniform decrease in temperature of the dry adiabatic lapse rate (DALR). (*See Q: How is the height of a cloud base determined? page 228.*) They have round and uneven tops because of the uneven decrease in the environmental lapse rate (ELR) temperature and different magnitudes in movement of the rising currents inside the cloud.

Note: If the ELR is high (i.e., the temperature of the surrounding air reduces more quickly with height gained than the rising DALR/SALR parcel of air), then this parcel of air will always be warmer than the surrounding (ELR) air. Therefore, it will be lighter and thus will keep rising. This air is said to be *unstable* and will produce cumiliform (i.e., heaped) clouds.

How are cloud types classified?

There are four main groups of clouds:

1. Curriform, or fibrous
2. Cumuliform, or heaped

3. Stratiform, or layered
4. Nimbus, or rain-bearing

These groups are further subdivided with the following prefixed names according to the level of their base above mean sea level (MSL):

1. Cirro, or high-level cloud: (cloud base > 16,500–20,000 ft
2. Alto, or medium-level cloud: cloud base > 6500 ft
3. No prefix, for low-level clouds: cloud base < 6500 ft

Note: The state or extent of any cloud formation depends on the stability of the air. (*See Q: Describe how clouds are formed, page 228.*)

If cumulus clouds were present in the morning, what would you expect later?

Cumulonimbus clouds (CBs).

Describe the formation of mountain (lenticular) clouds.

Airflow rises over mountains due to orographic uplift and cools adiabatically. If it cools below its dewpoint temperature, then the water vapor will condense out and form clouds, either as lenticular clouds, often on the hillside when there is a stable layer of air above the mountain, or as cumulus or even cumulonimbus clouds when there is unstable air above the mountain.

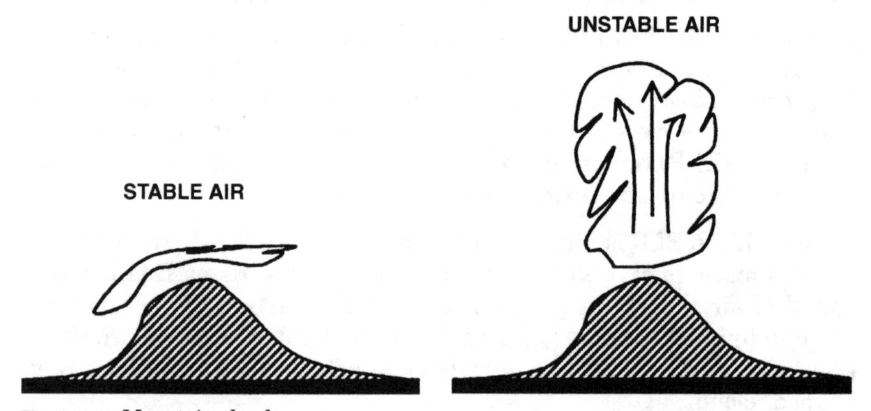

Figure 8.3 Mountain clouds.

Rotor, or roll, clouds, particularly common with lenticular cloud formation, also may form at a low level downstream of the mountain as a result of surface turbulence. (*See Q: What do lenticular clouds indicate? page 259.*)

What is mist and fog?

Mist and fog are simply parcels of low-level air in contact with the ground that have small suspended water droplets that have the effect of reducing visibility.

It is usual for mist to precede and to follow fog unless an already formed fog patch is blown in across an area, e.g., sea fog.

What are the different types of fog?

The most common different types of fog are

1. Radiation fog
2. Advection fog, including sea fog
3. Frontal fog, including hill fog

How are the different types of fog formed?

Fog in general is caused by a condensation process, due to a difference in temperature between the ground and the air next to it (radiation and advection fog) or between two interactive air masses (frontal fog), that, as a result, reduces the temperature of a parcel of air to its dew-point temperature.

1. *Radiation fog.* This requires the following conditions:
   a. *Cloudless night.* This allows the earth's surface to lose heat by radiation. This causes its water vapor to condense out in liquid form.
   b. *Moist air.* With a high relative humidity, which only requires a slight cooling to reach its dewpoint temperature.
   c. *Light winds.* Between 2 and 8 knots.

   Radiation fog occurs inland, especially in valleys and low-lying areas.
2. *Advection fog.* The term *advection* means heat transfer by the horizontal flow of air. Fog formed in this manner is called *advection fog* and can occur quite suddenly, day or night, land or sea, if the following conditions exist:
   a. A warm, moist air mass flowing across a significantly colder surface, which is cooled from below.
   b. Light to moderate winds that encourages the mixing of the lower levels to give a layer of fog.
   c. *Sea fog.* Sea smoke occurs in the reverse conditions; i.e., very cold air, often in an inversion, passes over a warmer sea that causes evaporating moisture to condense out into whispers of vapor.
3. *Frontal fog.* Usually forms in the cold air ahead of a warm occluded front as a prefrontal widespread fog. It forms due to the interaction of two air masses.

What is dew, and how is it formed?

Dew is a water cover on the earth's surface that is formed when the following conditions exist:

1. *Cloudless night.* This allows the earth's surface to lose heat by radiation and causes its water vapor to condense out as a water liquid.
2. *Moist air.* With a high relative humidity that only requires a slight cooling to reach its dewpoint temperature.
3. *Light winds.* Less than 2 knots.

Note: The conditions for dew to form are the same as for radiation fog except for the lower or nil wind.

What is frost, and how is it formed?

Frost is a frozen water cover on the earth's surface that is formed in the same manner as dew except that the earth's surface has a subzero temperature that causes the water droplets that have condensed out of the air to freeze on the ground. (*See Q: What is dew, and how is it formed? page 232.*)

For smog questions, see subchapter, "Visibility," page 265.

Storms and Precipitation

What is virga?

Virga is rain that falls from the base of a cloud but evaporates at a lower level in drier warmer air before it reaches the ground. This is a sign of a temperature inversion, which in turn is an indication of possible windshear. (*See Q: Where do you find windshear? page 254.*)

Virga is not really a form of precipitation because it does not reach the ground; however, it is generally considered to be precipitation, and it is important for pilots to recognize it because it can affect an aircraft's flight path.

Describe the formation of a thunderstorm.

Thunderstorms are associated with cumulonimbus clouds, and there may be several thunderstorm cells within a single cloud. Therefore, we first have to examine how a cumulonimbus cloud starts to form.

Four conditions are required for a cumulonimbus cloud to develop:

1. A high moisture content in the air.
2. A trigger lifting action (or catalyst) to cause a parcel of air to start rising. The four main lifting actions are
   a. Convection
   b. Turbulence
   c. Frontal
   d. Orographic
3. Adiabatic cooling of the rising air.
4. A highly unstable atmosphere so that once the air starts to rise, it will continue rising. Effectively, the environmental lapse rate (ELR) must be greater than the saturated adiabatic lapse rate (SALR) for over 10,000 ft. (*See Q: Describe how clouds are formed? page 228.*)

Next, we have to examine the life cycle of the cumulonimbus cloud and its associated thunderstorm. This life cycle can be divided into three phases: the developing stage, the mature stage, and the decaying stage.

1. *Developing stage.* During the development of the cumulonimbus cloud, updrafts move air aloft, allowing condensation to take place throughout the ascent of the convective currents.
2. *Mature stage.* During this stage, water drops start to fall through the cloud, drawing air down with them. Although it is dependent on the shape of the storm and the prevailing wind gradient, this downdraft is often in the middle of the cloud/storm, surrounded on all sides by strong continuing updrafts, which are providing further fuel for the storm. During this stage, downdrafts can reach 3000 ft/min, and updrafts can reach 5000 to 6000 ft/min. The mature phase of a cumulonimbus cloud is also the most hazardous stage of its thunderstorms. In short, the dangers include
   a. Torrential rain
   b. Hail
   c. Severe turbulence
   d. Severe icing
   e. Windshear and microbursts
   f. Lightning
3. *Decaying stage.* This is the final stage of the cumulonimbus cloud. It starts with the end of the thunderstorm, which is marked by the end of continuous rain and the start of sporadic showers, sometimes as virga due to a temperature inversion beneath the cloud base, which can still cause a marked windshear. At the higher levels it may take on the familiar anvil shape as upper winds spread out under the tropopause. An anvil can have marked downward vertical currents beneath it, which cause a strong windshear that also should be avoided.

How are thunderstorms a hazard to aviation?

Thunderstorms can produce the following hazards to all aircraft types:

1. Severe windshear, which can cause
   a. Handling problems
   b. Flight path deviations, especially vertically
   c. Loss of airspeed
   d. Possible structural damage
   (*See Q: What is windshear? page 254.*)
2. Severe turbulence, which can cause
   a. Possible loss of control
   b. Possible structural damage
   (*See Q: What is turbulence? page 253.*)
3. Severe icing, especially clear ice formed from supercooled water droplets (SWDs) striking a surface with a subzero temperature. (*See Qs: What are SWDs? page 262; What hazards to aviation does icing cause? page 262.*)
4. Airframe structural damage from hail
5. Reduced visibility
6. Lightning strikes, which can cause damage to the electrical system
7. Radio communication and navigation interference from static electricity present in the thunderstorm

These hazards exist inside, under, and for some distance around a thunderstorm's associated cumulonimbus cloud. Therefore, cumulonimbus clouds (and thunderstorms) should be avoided by a minimum of 10 nautical miles and in severe conditions (i.e., the mature stage of a cumulonimbus cloud) by at least 20 nautical miles.

When is lightning most likely to occur?

Lightning is most likely to occur when the outside air temperature (OAT) is +10°C to −10°C and within or close to a thunderstorm associated with a mature-stage cumulonimbus cloud.

Winds

What is wind?

Wind is the horizontal movement of air in the atmosphere that is driven initially by a difference in pressure between two places, and then it can be influenced further by a number of factors, including the earth's rotation forces, temperature, and surface friction.

How is wind described or expressed?

Wind is expressed in terms of direction and strength.

1. *Wind direction.* This is the direction *from which* the wind blows and is expressed in degrees measured clockwise from true north, with two main exceptions when the wind direction is measured from magnetic north:
   a. The reported wind given from the tower, either air traffic control (ATC) or Automatic Terminal Information Service (ATIS), is expressed in degrees magnetic so that the runway direction and reported wind are relative to each other. This is extremely important when taking off or landing.
   b. The upper winds used for airways planning are expressed in degrees magnetic so that the airway direction and reported wind are relative to each other.
2. *Wind strength.* This is measured and expressed in nautical knots.

What is wind velocity?

Wind velocity relates to the wind's direction and strength. It is usually written in the following form: 180/25; i.e., a wind blowing from 180° at a strength of 25 knots.

Describe a veering and backing wind.

A wind is said to be *veering,* or is said to have *veered,* when it changes its direction in a clockwise direction, e.g., 100/10 to 160/10. A wind is said to be *backing,* or is said to have *backed,* when it changes its direction in an counterclockwise direction, e.g., 160/10 to 100/10. (*See Q: Describe the characteristics of a surface wind, page 238.*)

What is buys ballot's law?

Buys ballot's law is if you stand with your back to the wind in the northern hemisphere, the low pressure (temperature) will be on your left.

Note: Conversely, in the southern hemisphere, the low pressure (temperature) will be on your right.

What is the pressure gradient force?

A pressure gradient force is a natural force generated by a difference in pressure across a horizontal distance; i.e., gradient between two places. It acts at right angles to the isobars and is usually responsible for starting the movement of a parcel of air from an area of high pressure to an area of low pressure. (*See Q: What is an isobar? page 221.*)

What is the coriolis force (or geostrophic force)?

The coriolis (geostrophic) force is an apparent force that acts on a parcel of air that is moving over the rotating earth's surface. This means that the air does not flow simply from a high- to low-pressure system but is deflected to the left or right according to which hemisphere you are in. This is known as the *coriolis effect.* The coriolis force is a product of the earth's rotational properties.

Note: In the northern hemisphere, the coriolis force deflects the airflow to the right (i.e., as a westerly wind), and in the southern hemisphere, the airflow is deflected to the left.

What is the geostrophic wind?

The geostrophic wind is the balanced flow of air from a westerly direction that is parallel to straight isobars with a low-pressure system to its left, i.e., buys ballot's law, and at a strength directly proportional to the spacing of the isobars (i.e., pressure gradient). It is usually found at low to medium heights from approximately 2000 ft and above.

The geostrophic wind is created when the two forces of pressure gradient force and coriolis (geostrophic) force are balanced.

What is a gradient wind?

The gradient wind is the resulting wind that blows around the curved isobars common to circular low- or high-pressure patterns. It is usually found at low to medium heights from approximately 2000 ft and above.

Note: The geostrophic wind assumes straight west to east isobars. (*See Q: What is the geostrophic wind? page 236.*)

Describe upper winds.

Upper winds are determined by the thermal gradient. A difference in temperature between two columns of air will cause a pressure difference at height even if both columns of air have the same sea level pressure. This pressure difference creates a wind (parallel to the isobars) at altitude that is different from the wind experienced at sea level, even if no wind was present at sea level.

The vector sum of the isotherm thermal wind component and the surface and upper isobar pressure-driven geostrophic (or gradient) wind produces the direction and speed of the upper wind.

In the northern hemisphere, the thermal gradient is generally north-south (north cold and south warm), and therefore, the upper winds generally are westerly in direction (i.e., from the west), with the highest wind speed where the thermal gradient is greatest, e.g., jetstreams.

Note: However, there is a light easterly wind over the thermal equator that can extend all around the globe at certain times of the year and can reach speeds of up to 70 knots.

What is a thermal wind, and how is it generated?

Thermal winds (of quite different direction and strength to the low- and medium-level geostrophic and gradient winds) are generated by a difference in temperature (thermal gradient) between two columns of air over large areas and at great upper heights.

The direction of a thermal wind is parallel to the isotherms. (*See Q: What is an isotherm? page 221.*) That is, the thermal wind direction is such that if you stand with your back to the thermal component in the northern hemisphere, the low temperature will be to your left. The strength or speed of the thermal wind is directly proportional to the temperature gradient (i.e., the spacing of the isotherms) between the two columns of air.

What is a jetstream?

Jetstreams are simply narrow bands of high-speed upper thermal winds at very high altitudes. (*See Q: What is a thermal wind, and how is it generated? page 237.*)

The official definition of a jetstream is a strong, narrow current on a quasi-horizontal axis in the upper tropopause or stratosphere characterized by strong vertical and/or lateral windshear (CAT). The wind speed must be greater than 60 knots for a wind to be classified as a jetstream.

Jetstreams typically are 1500 nautical miles long, 200 nautical miles wide, and 12,000 ft deep, and their speed is directly proportional to the thermal gradient; i.e., the greater the thermal gradient, the greater is the speed of the jetstream.

Where do you find jetstreams?

Jetstreams are driven by thermal gradients and therefore are found wherever the thermal gradient is high enough. There are two bands of rapid temperature changes (i.e., high/maximum thermal gradient) in each hemisphere that are marked enough to produce a jetstream. They are

1. At the polar front around 60 degrees of latitude, where the polar air meets the subtropical air. This is a *polar front jetstream,* and is the most marked thermal gradient to be found, especially when it is over land in the winter.

2. At the intertropical front, where the subtropical air meets the tropical air. This is known as the *intertropical front jetstream.*

The jetstream exists just below the tropopause in the warm air of a pressure system; i.e., in the subtropical air at the polar front and in the tropical air at the intertropical front, but appears on the surface chart to be in the cold sector. This is so because of the slope of the front with height.

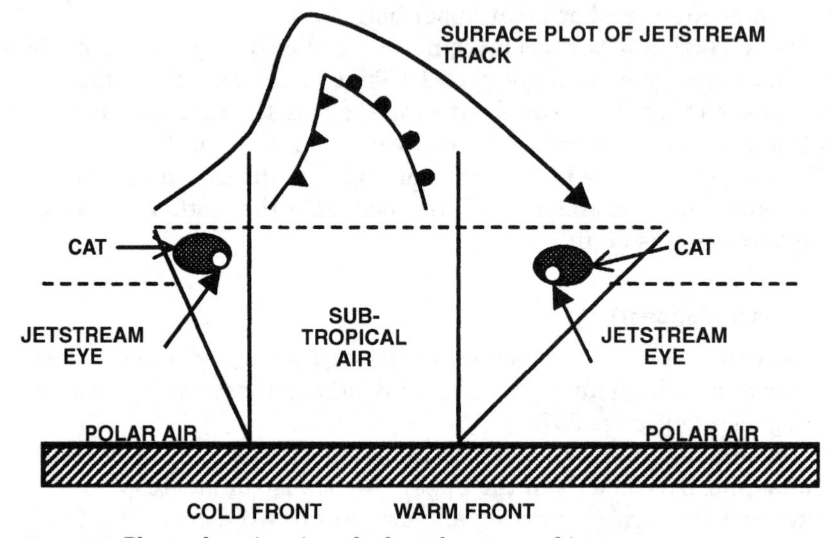

Figure 8.4 Plan and section view of a frontal system and jetstream.

The jet moves with the front (i.e., south in the winter), and its direction is not always westerly because the jet follows the pressure system. In fact, the polar front jetstream can blow from 190 to 350 degrees around the polar front. However, its overall direction generally is regarded as being from west to east.

The maximum windshear/clear air turbulence (CAT) associated with a jetstream can be found level with or just above the jet core in the warm air but on the cold polar air side of the jet.

Describe the characteristics of a surface wind.

At the surface, the wind weakens in strength (speed) and backs in direction in the northern hemisphere (veers in direction in the southern hemisphere).

The wind speed reduces near the surface compared with the free air geostrophic/gradient wind at 2000 ft due to the friction forces between the moving air and the ground.

Describe the diurnal variation of the surface wind.

The diurnal variation (time of day) affects the degree to which the underlying trend of the surface wind is altered, i.e., to weaken in strength and back in direction (northern hemisphere) compared with the free air gradient wind (at approximately 2000 ft). (*See Q: Describe the characteristics of a surface wind, page 238.*)

By day. The day surface wind loses less of its strength/speed and therefore backs only slightly compared with the free air gradient wind or is a stronger wind that has veered compared with the night surface wind.

By night. The night surface wind will drop in strength (speed) and back in direction significantly compared with the free air gradient wind and to a slightly lesser extent compared with the daytime surface wind.

Therefore, the surface wind by day will resemble the gradient wind more closely than the surface wind at night. (*See Q: What are the approximate changes in the surface wind direction and speed compared with the free air 2000-ft gradient wind? page 239.*)

How does the wind (direction and speed) change with height?

In general, the wind in the northern hemisphere increases in speed and veers in direction with an increase of height. And in the southern hemisphere, the wind increases in speed and backs in direction with an increase of height.

What are the approximate changes in the surface wind direction and speed compared with the free air 2000-ft gradient wind?

The surface wind reduces in speed and in the northern hemisphere backs in direction compared with the free air gradient wind at 2000 ft. (*See Q: Describe the characteristics of a surface wind, page 238.*) However, the degree to which it changes direction and speed is a function of

1. Diurnal variation (i.e., night or daytime). (*See Q: Describe the diurnal variation of the surface wind, page 239.*)
2. Surface variation (i.e., land or sea).

The following table gives a good approximation of the variation of the surface wind compared with the free air gradient wind (at approximately 2000 ft) above the surface layer of turbulence, considering both diurnal and surface variations, in the northern hemisphere.

	Day		Night	
	Direction	Speed	Direction	Speed
Land	−20°	×0.5	−40°	×0.3
Sea	−15°	×0.7	−15°	×0.7

Note: In the southern hemisphere, the surface wind veers in direction compared with the free air gradient wind.

Why is the surface wind important to pilots?

The surface wind is important to pilots because it relates directly, in terms of both its speed and its direction, to the effect it has on their aircraft during takeoffs and landings.

What are trade winds?

Trade winds are steady and predictable surface winds that rarely exceed 15 knots at the surface but can extend up to 10,000 ft. They blow from the subtropical highs into the equatorial low (ITCZ), i.e., blowing from the northeast in the northern hemisphere and from the southeast in the southern hemisphere. (*See Q: What is the ITCZ? page 266.*)

Explain land/sea breezes, especially in connection with coastal airports.

Land/sea breezes are a product of surface heating and atmospheric convection currents that produce small local airflow circulation cells. Land/sea breezes are particularly important to aviators in connection with coastal airports because they can determine the wind direction at a local level, which may be considerably different from the general wind direction. In addition, they also may cause some windshear and/or turbulence as an airplane passes from one body of air to another.

Sea breezes. These occur during the daytime, usually between midday and late afternoon on hot sunny days, when the land heats up quicker than the sea. The air above the land becomes warmer and rises, causing the pressure in the warmer air column over the land to be greater than that at a similar height over the sea. Therefore, a pressure difference is born at height, i.e., 2000 ft, that produces an airflow at height from the land to the sea. This induces an opposite pressure difference at the surface; namely, the outflow of air at height cools, gets denser and heavier, and therefore descends over the sea, creating a high surface pressure. This induces a flow of air at the surface from the sea to the land (which has a low surface pressure due to the rising convective current air), which is the sea breeze.

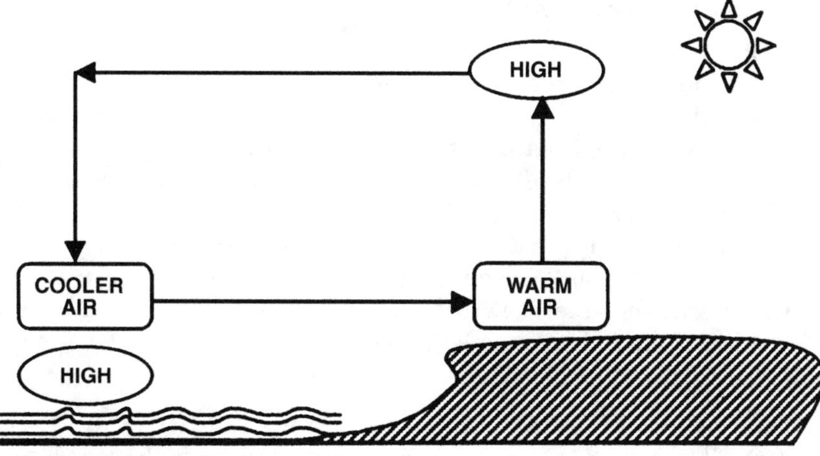

Figure 8.5 Sea breeze by day.

Land breezes. These occur at night when the land cools quicker than the sea, causing the air above it to cool and subside. The air over the sea, however, retains its heat better and becomes the warmer of the two columns of air, and therefore, its air rises. The pressure in the warmer air column above the sea is greater than that over the land at height, so the air flows at height from the sea to the land. This induces an opposite pressure difference at the surface between the land and the sea; namely, the outflow of air at height cools, gets denser and heavier, and therefore descends over the land, creating a high surface pressure. This induces a flow of air at the surface from the land to the sea (which has a low surface pressure due to the rising warmer air), which is the offshore land breeze.

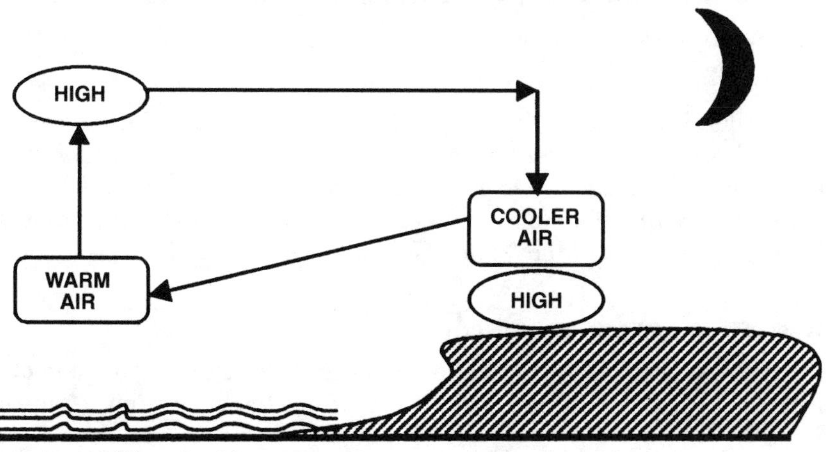

Figure 8.6 Offshore land breeze by night.

What is the fohn wind effect?

The fohn wind effect is when air is cooled as it is forced to rise over high ground, first at the dry adiabatic lapse rate (DALR) and then, after the dewpoint/condensation level is reached, at the saturated adiabatic lapse rate (SALR). After the condensation level, clouds will form, usually on the hillside, and moisture will be lost, falling as either rain or dew.

Then, as the air descends over the far side of the hill, it has a lower water content, and so the condensation level is higher. A longer period of warming at the greater DALR means that the air on the far (downwind) side of the hills is warmer and dryer than it was on the upwind side of the hill.

Note: As a rough guide, the air temperature can rise by 1.5°C per 1000 ft of hill, which of course is the difference between the DALR and the SALR.

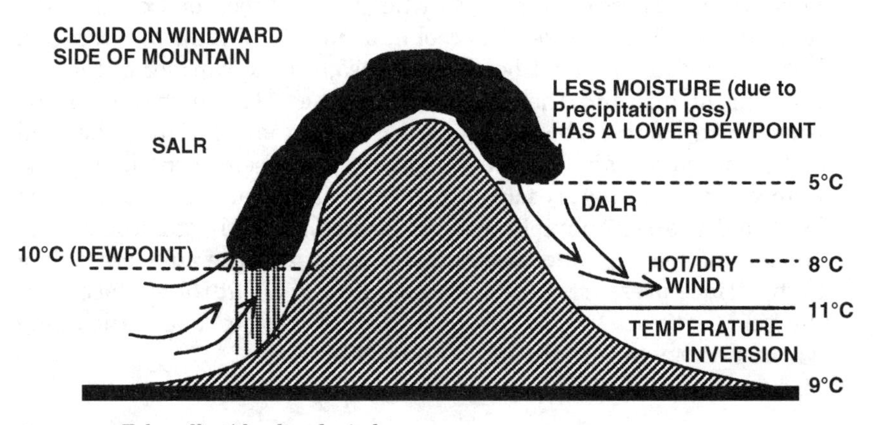

Figure 8.7 Fohn effect/cloud and wind.

Therefore, the fohn effect (i.e., descending air) can influence the wind direction and strength and in some cases even generate a wind that blows down the far/lee side (downwind) of the hill. This is known as the *fohn wind.* The fohn wind effect can be seen in various places, including the Rocky Mountains (Chinook winds) and the Alps.

What is a katabatic/anabatic wind?

A katabatic wind is a local valley wind that flows down the side of a hill. An anabatic wind is a local valley wind that flows up the side of a hill.

Air Masses/Pressure Systems/Fronts

What is an airmass?

An airmass is a large parcel of air with fairly similar temperature and humidity (moisture content) properties throughout. For this definition it follows that the environmental lapse rate (ELR) (*see Q: What is ELR? page 226*) will be fairly constant throughout the airmass, and all the air will behave in the same way when lifted, heated, or cooled.

What is a pressure system?

A pressure system is a circulating airmass that is classified as either a low or high, which relates to the direction of pressure change toward the center of the airmass at the surface, i.e., gets lower or higher.

A low-pressure system typically will have more than one airmass, e.g., a warm airmass incorporated into a cold unstable airmass, with fronts in between. (*See Q: What is a front? page 249.*) On the other hand, a high-pressure system is typically made up of a single stable airmass. (*See Qs: Describe the characteristics of a low-pressure system, page 243; Describe the characteristics of a high-pressure system, page 246.*)

Describe the characteristics of a low-pressure system.

A low-pressure system, which is also known as a *depression,* has the following characteristics:

1. *Pressure gradient.* In a low-pressure system, the surface barometric air pressure rises as you move away from its center; in other words, its pressure drops as you move toward its center.
2. *Airflow pattern.* The airflow pattern of a depression can be categorized as follows:
   a. Convergence (inflow) at the lower layers
   b. Rising air at the center
   c. Divergence (outflow) in the upper layers
3. *Wind direction.* (*See Q: What is the wind direction around a low-pressure system? page 244.*)
4. *Airmass.* A low-pressure system usually will be made up of at least two different air masses, i.e., a cold and a warm air mass, with cold and warm fronts.
5. *Movement.* Low-pressure systems generally are more intense than highs because they are more concentrated in terms of area and have a stronger pressure gradient (pressure change with distance). Therefore, lows move faster across the surface of the earth and tend to have a shorter life span than highs.

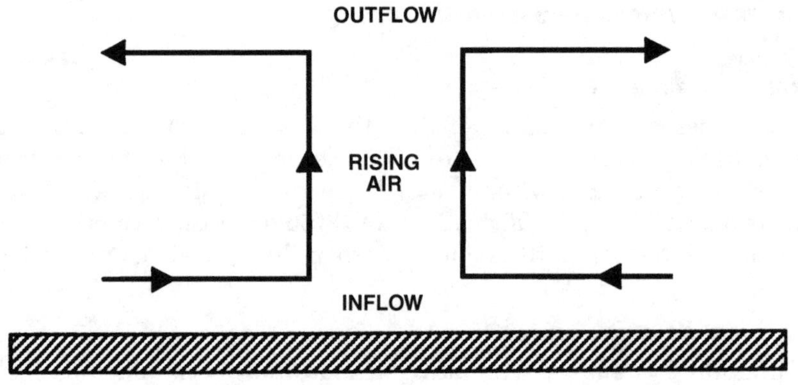

Figure 8.8 Airflow pattern in a depression.

6. *Weather.* (*See Qs: Describe the weather associated with a depression, page 244; What are the different types of depressions? page 244.*)

What is the wind direction around a low-pressure system?

The wind circulates counterclockwise around a low-pressure system in the northern hemisphere and clockwise in the southern hemisphere.

Note: Flying toward a low in the northern hemisphere, an aircraft will experience right (starboard) drift. (*See Q: What drift would you experience when flying from a high- to a low-pressure system? page 245.*)

What are the different types of depressions?

The different types of depressions (low-pressure systems) are developed and classified according to their trigger phenomenon. Depressions are classified as follows:

1. Frontal depression
2. Thermal depression
3. Tropical storm depressions
4. Orographic depression

Describe the weather associated with a depression.

In general, the weather associated with a low-pressure system is as follows:

1. Cloud formation and related weather are present, i.e., precipitation, etc. This is due to the adiabatic cooling experienced by the

ascending air in a depression. Any instability in the rising air may lead to the vertical development of cumuliform clouds accompanied by rain showers.

2. Visibility may be good (except in the rain showers). This is so because the vertical motion will tend to carry away most of the particles suspended in the air.
3. Moderate to strong winds are present. This is so because a low-pressure system typically has a marked pressure gradient, which represents the strength of the wind.
4. Frontal weather is present because fronts are associated with low-pressure systems; thus frontal weather is normally the most prominent weather associated with a depression. (*See Q: Describe the characteristics and weather common to the passage of a warm/cold front, pages 250 and 252.*)

What is a trough?

A trough is a V-shaped extension of a low-pressure system. Air flows into a trough (convergence) and rises. If the air is unstable, weather similar to that found in a depression or a cold front will occur, e.g., cumuliform clouds, cumulonimbus clouds, and thunderstorm activity.

What drift would you experience when flying from a high- to a low-pressure system?

In the northern hemisphere, you would experience a drift to the right (starboard) when flying from a high- to a low-pressure system. This is so because the wind circulates counterclockwise around a low in the northern hemisphere, and therefore, flying toward the center of a low-pressure system, you have a wind from the left (port) causing you to drift to the right (starboard). (*See Qs: Describe the characteristics of a low-pressure system, page 243; Describe the characteristics of a high-pressure system, page 246.*)

Note: It is worth remembering the high to low mnemonic: High to low, beware below. If you are flying from a high to a low temperature or pressure area, your altimeter will overread. (*See Q: What pressure altitude error is commonly experienced? page 117.*)

In the southern hemisphere, when flying from a high- to a low-pressure system, the opposite is true; namely, you would experience a drift to the left (port). This is so because in the southern hemisphere the wind circulates clockwise around a low and counterclockwise around a high, and therefore, the airflow/wind experience will be coming from the right, causing you to drift to the left (port).

What are tropical revolving storms (TRSs)?

(*See page 266.*)

Describe the characteristics of a high-pressure system.

A high-pressure system, which is also known as an *anticyclone,* has the following characteristics:

1. *Pressure gradient.* In a high-pressure system, the surface barometric air pressure rises as you move toward its center; in other words, its pressure drops as you move away from its center. The pressure gradient (which is depicted by the closeness of the isobars) relates to the wind speed; i.e., a weak pressure gradient = wide isobar spacing = a weak wind.
2. *Airflow pattern.* The airflow pattern of an anticyclone can be categorized as follows:
   a. Convergence (inflow) at the upper layers
   b. Subsiding air at the center
   c. Divergence (outflow) at the lower layers

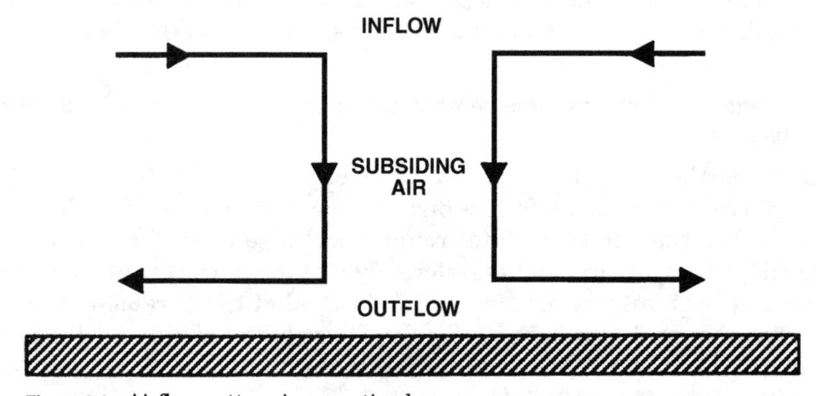

Figure 8.9 Airflow pattern in an anticyclone.

3. *Wind direction.* (*See Q: What is the wind direction around a high-pressure system? page 247.*)
4. *Airmass.* A high-pressure system usually will be made up of only one air mass and therefore will have no defined fronts.
5. *Movement.* High-pressure systems generally are greater in extent but with a weaker pressure gradient (change of pressure with distance) and therefore are slower moving, more persistent, and last longer compared with a low-pressure system.

6. *Weather.* (*See Q: Describe the weather associated with a high-pressure system, page 247.*)

What is the wind direction around a high-pressure system?

The wind circulates clockwise around a high-pressure system in the northern hemisphere and counterclockwise in the southern hemisphere.

Note: Flying toward a high in the northern hemisphere, an aircraft will experience left (port) drift. (*See Q: What drift would you experience when flying from a high- to a low-pressure system? page 245.*)

However, it should be remembered that above a surface high there is an upper low system. (*See Q: Describe the characteristics of a high-pressure system, page 246.*) Therefore, the wind direction will reverse at height in a high-pressure system.

Describe the weather associated with a high-pressure system.

In general, the weather associated with a high-pressure system is as follows:

1. *Clear upper skies with little or no low-level cloud formation and precipitation.* This is so because the descending stable air (subsiding air is very stable) in a high-pressure system is warmed as it descends. Therefore, the dewpoint temperature is increased, and the relative humidity is reduced. Thus cloud formation and any associated precipitation will tend to disperse.
2. *Light winds.* This is so because a high-pressure system has a weak pressure gradient, and its wind strength is represented by its pressure gradient.
3. *Possible poor visibility at low levels.* This is so because the descending air may bring airborne particles from the upper levels down to the lower levels that we as pilots fly in or may warm sufficiently to create an inversion layer as a result of the upper descending air warming to a greater temperature than the lower surface turbulent layer. This phenomenon may cause stratiform clouds to form and/or trap airborne particles, dust, smoke, and moisture (fog) beneath the inversion layer.

What is a ridge?

A ridge is a U-shaped extension to a high-pressure system.

Stable air subsides into a ridge, similar to a high-pressure system, and therefore, weather similar to that found in an anticyclone will occur in a ridge. (*See Q: Describe the weather associated with a high-pressure system, page 247.*)

What drift would you experience when flying from a low- to a high-pressure system?

In the northern hemisphere you would experience a drift to the left (port) when flying from a low- to a high-pressure system. This is so because the wind circulates clockwise around a high in the northern hemisphere, and therefore, flying toward the center of a high-pressure system, you would have a wind from the right (starboard) causing you to drift to the left (port).

In the southern hemisphere when flying from a low- to a high-pressure system, the opposite is true; namely, you would experience a drift to the right (starboard). This is so because in the southern hemisphere the wind circulates clockwise around a low and counterclockwise around a high, and therefore, the airflow/wind experienced will be coming from the left, causing you to drift to the right (starboard).

Describe the airflow between a low- and a high-pressure system.

The airflow between a low- and a high-pressure systems is pressure-driven. As a random starting point, at the surface of a high-pressure system the air flows outward (divergence) at low levels into the center of a surface low-pressure system (convergence). Here it rises to an upper-level high and flows outward (divergence) and into an upper-level low (convergence), where it subsides down to where it started, our surface high pressure. (*See Qs: Describe the characteristics of a high-pressure system, page 246; Describe the characteristics of a low-pressure system, page 243.*)

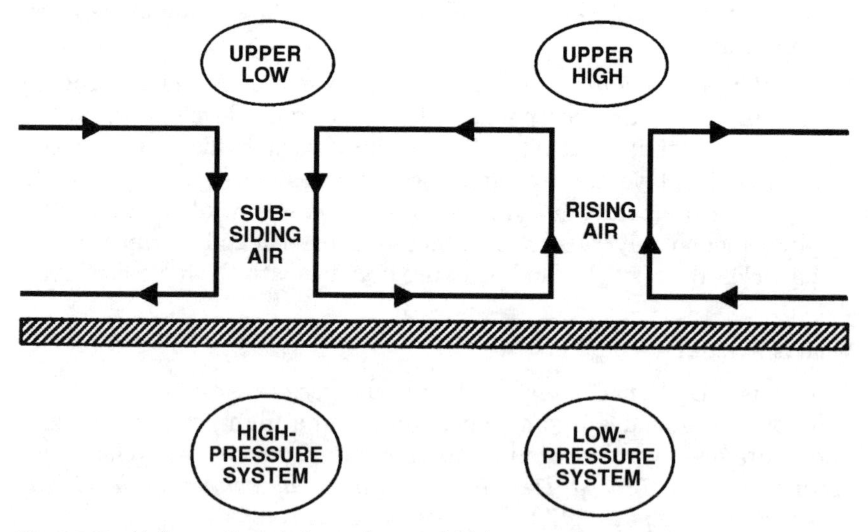

Figure 8.10 Airflow pattern between low- and high-pressure systems.

How does a change in pressure affect an aircraft's flight level?

A change in atmospheric pressure causes a barometric pressure error in the altimeter. (*See Q: What pressure altitude error is commonly experienced? page 117.*) Therefore, when flying from high to low (pressure), beware below because your actual flight level is lower than your altimeter indicated level. In other words, the altimeter overreads. Conversely, the opposite is true when flying from low- to high-pressure areas. This high- to low-mnemonic applies equally to temperature values as it does pressure. (*See Q: How does a change in air temperature affect an aircraft's flight level? page 225.*)

What is a front?

A front is a boundary between two different air masses.

Because air masses have different characteristics depending on their (1) origin, (2) path over the earth's surface, and (3) diverging or converging air (*see Q: What is an air mass? page 243*), they usually give rise to distinct divisions between adjacent air masses, which are known as *fronts*. There are two basic types of fronts, cold and warm. (*See Qs: What is a warm front? page 249; What is a cold front? page 251.*) Fronts normally are found in low-pressure systems because a depression is usually made up of two different air masses. (*See Q: Describe the characteristics of a low-pressure system, page 243.*)

What is frontal activity?

Frontal activity describes the interaction between at least two air masses as one air mass replaces another.

How are frontal depressions developed?

Frontal depressions (low-pressure frontal system) develop when two different air masses meet (touch) but do not mix together because one of them is warmer and less dense than the other.

What is a warm front?

A warm front is the boundary produced between two air masses (i.e., warm air behind cold air), where the warmer, less dense air mass rises up and replaces at altitude (slides over) the colder air mass at the surface.

In a warm front, the frontal air at altitude is actually well ahead of its depicted position on a weather chart.

Note: A front is shown as a line on a weather chart that represents the front's surface position. A warm front is represented by half circles along the front line.

In fact, the slope of the warm front is typically 1:150, which is much flatter than a cold front. Therefore, the top of a warm front, which is usually characterized by cirrus clouds, could be up to 600 nautical miles ahead of the surface front, and altostratus rain-bearing clouds could be up to 200 nautical miles ahead of it. A warm front is at the leading edge of the warm (air mass) sector and ahead of the cold front in a depression (low-pressure system).

Describe the characteristics and weather common to the passage of a warm front.

The passage of a warm front has the following general characteristics and associated weather:

1. As the warm front approaches:
   a. A lowering of the cloud base is experienced. This is represented by cirrus clouds giving way to cirrostratus clouds giving way to altostratus clouds with possible virga rain giving way to nimbostratus clouds with increased rainfall.
   b. Poor visibility is experienced. This is so because with the passage of a warm front, visibility is reduced due to the increase in low-level clouds and the more consistent rainfall and possibly rain ice.
   c. The atmospheric pressure usually will fall as a warm front approaches.

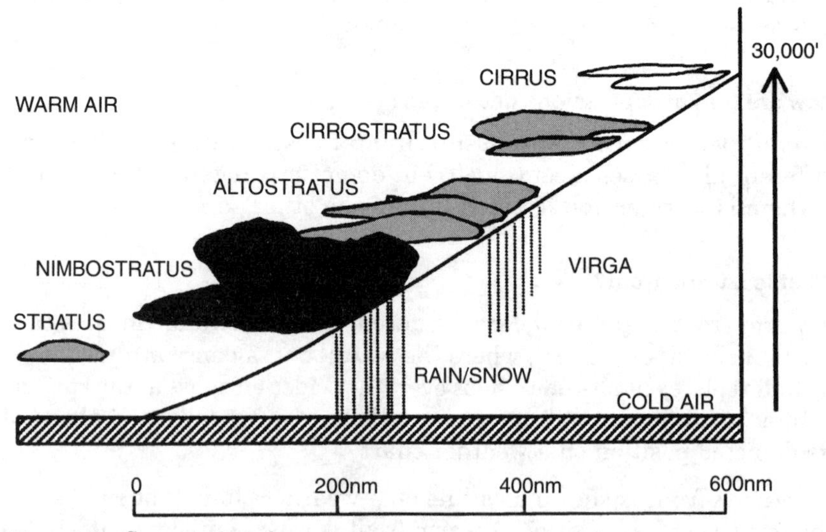

Figure 8.11 Cross section of a warm front.

2. As the warm front passes:
   a. A rise in air temperature. This is due to the arrival of the warm sector air mass.
   b. Low-level cloud base or fog.
   c. Wind veers in direction in the northern hemisphere and backs in the southern hemisphere.
   d. The atmospheric pressure usually will stop falling and may even rise.
   e. Good visibility. In general, the visibility will be good, especially above the low-level clouds, although obviously the visibility will be poor in the cloud or fog.

What is a cold front?

A cold front is the boundary between two air masses (i.e., cold air behind warm air), where the colder, denser air mass undercuts and replaces the warmer preceding air mass from the surface upward (slides under).

The cold front position on a weather chart rather accurately displays the front's actual position because the upper altitude position of the front is virtually overhead its surface position.

Note: A front is shown as a line on a weather chart that represents the front's surface position. Barbs along the front line represent a cold front.

This is so because the slope of the cold front is relatively steep, approximately 1:50. Therefore, there is little warning of the approach of a cold front, which typically contains cumulus and even cumulonimbus clouds with associated thunderstorms, covering a band of only 30 to 50 nautical miles.

A cold front is at the trailing edge of the warm (air mass) sector and generally behind the warm front in a depression (low-pressure system). It moves quicker than the warm front ahead of it, with its cooler frontal air at altitude lagging slightly behind the cold air at the surface.

Why does a warm front move slower than a cold front?

A warm front moves slower than a cold front because of the tendency of the warm air (in the warm front) to rise up and slide over the cold, more resistant, denser air (mass) in front of it. Therefore, the force (geostrophic wind component) moving a warm front is used to move the warm air upward as well as horizontally forward. Hence the speed of a warm front is only two-thirds the geostrophic wind component. However, the denser cold air in the cold front meets very little resistance from the warm, less dense sector air in front of it, and therefore, its driving force is used solely to move it horizontally forward at roughly the full geostrophic wind component.

Describe the characteristics and weather common to the passage of a cold front.

(*See Figure 8.12.*)
The passage of a cold front has the following general characteristics and associated weather:

1. As the cold front approaches:
   a. Cumulus and even cumulonimbus cloud cover is experienced. This results in the following severe weather changes:
      (1) Heavy rain
      (2) Thunderstorms
      (3) Turbulence (possibly severe)
      (4) Windshear
   b. Poor visibility is experienced. This is so because with the passage of a cold front, visibility is reduced due to the increase in heavy clouds through a large vertical band and the heavy, consistent rainfall.
   c. The atmospheric pressure usually will fall as the cold front approaches.
2. As the cold front passes:
   a. A sudden drop in air temperature is seen. This is due to the arrival of the cold sector airmass.
   b. Clear skies with isolated cumulus clouds are seen.
   c. There is good visibility, except within the isolated cumulus clouds.
   d. Wind veers in direction in the northern hemisphere and backs in the southern hemisphere.
   e. The atmospheric pressure will stop falling and may even rise.

What is an occluded front?

An occluded front is a combination of both a cold and a warm front. (*See Qs: What is a warm front? page 249; What is a cold front? page 251.*)

Note: A front is shown as a line on a weather chart that represents the front's surface position. An occluded front is represented by a combination of semicircles (warm front symbol) and barbs (cold front symbol) along the front line.

An occlusion occurs because the cold front moves faster than the warm front, and therefore, the cold front inevitably will catch up with the warm front. This occurs first near the center of the depression, where the original distance between the two fronts is the shortest.

The characteristic weather of the two fronts also becomes combined. However, because temperatures between the cold air behind the warm sector and the cold air in front of the warm sector are never the same, the type of occlusion will be either a warm or cold occlusion.

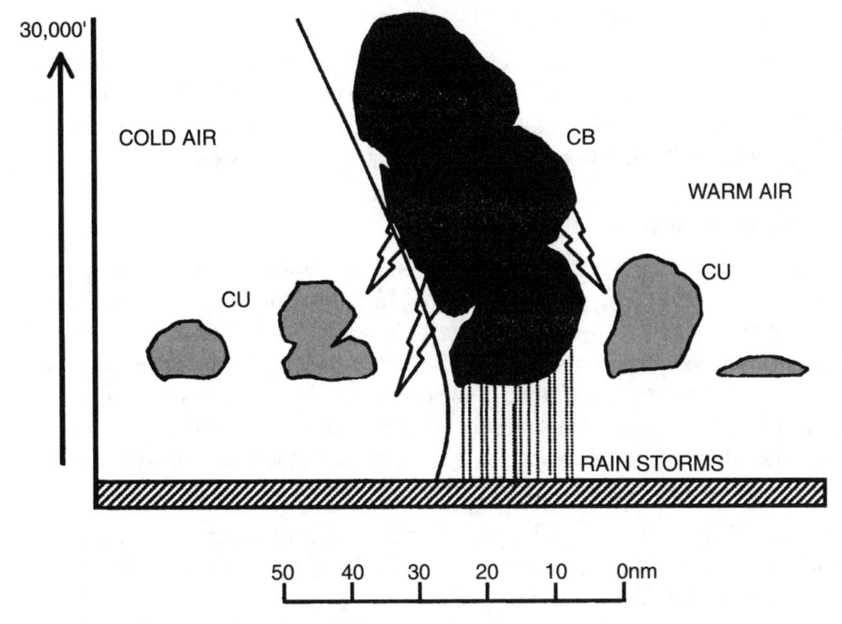

Figure 8.12 Cross section of a cold front.

Turbulence

What is turbulence?

Turbulence is the eddy motions in the atmosphere, which vary with time, from place to place, and in magnitude.

Some form of turbulence is always present in the atmosphere, with severe or even moderate turbulence at best being uncomfortable and at worst capable of overstressing some airframe types.

Turbulence is mainly considered to be vertical gusts. The two main forms of vertical turbulence are

1. Convection turbulence, caused by solar radiation heating the ground and producing rising thermal currents
2. Obstruction and orographic (terrain-generated) turbulence

The other main forms of turbulence (considered to be horizontal) are

1. Jetstreams
2. Wake turbulence (*See Q: What is wake turbulence? page 260.*)

In addition, other (horizontal) forms of less dynamic turbulence are frontal, temperature, and any general wind direction and/or speed changes. (*See Q: What is CAT, and give examples? page 259.*)

What causes surface turbulence?

Surface turbulence is caused by the surface wind being blown over and around surface obstacles, such as hills, trees, and buildings. This causes the surface wind to form turbulent eddies, the size of which depends on the size of both the obstruction and the wind speed.

What is windshear?

Windshear is any variation of wind speed and/or direction from place to place, including updrafts and downdrafts. The stronger the change and/or the shorter the distance within which it occurs, the greater is the windshear. However, in practical terms, only a wind change of a magnitude that causes turbulence or a loss of energy disturbance to an aircraft's flight path is generally considered to be windshear. It should be remembered that windshear is a complex subject and is still not fully understood. (*See Q: Where do you find windshear? page 254.*)

Note: Windshear can affect the flight path and airspeed of an aircraft and can be a hazard to aviation. (*See Q: How does windshear affect an aircraft? page 255.*)

Where do you find windshear?

Most forms of windshear are found at low levels, i.e., below 3000 ft. Therefore, the term *low-level windshear* is used to specify the windshear along a final approach path prior to landing, along a runway, and along a takeoff and initial climb-out flight path.

Low-level windshear, i.e., near the ground, generally is present to some extent as an aircraft approaches the ground because of the difference of speed and/or direction of the surface wind compared with the wind at altitude. This is critical in terms of aircraft safety because the loss of energy caused by these changes can easily lead to the aircraft losing altitude and/or stalling. (*See Q: How would you fly an approach if you suspect windshear? page 319.*) Low-level windshear includes

1. Clear air turbulence (CAT). (*See Q: What is CAT, and give examples? page 259.*)
2. Frontal passage
3. Microburst and thunderstorm gusts. (*See Q: What is a microburst? page 256.*)

Medium- and high-level windshear also can be experienced at high or medium altitudes and include

1. CAT in the form of jetstreams. (*See Qs: What is a jetstream? page 237; Where do you find jetstreams? page 237.*)
2. Frontal passage (e.g., a low-level frontal passage)

How is windshear detected?

Windshear is detected by identifying a difference in wind (direction and/or speed) and/or temperature between two places or identifying certain weather phenomenon, e.g., cumulonimbus clouds, that give rise to possible windshear conditions from weather reports. This can be done in the following ways:

1. Pilot appreciation of differences in reported wind and/or temperature between two places and the location of certain weather and terrain phenomenon. For example, the following phenomena should alert the pilot instinctively of the possibility of windshear along his or her's flight path:
   a. Cumulonimbus clouds in the general area (microbursts and/or thunderstorm gusts)
   b. Heavy rain or even thunderstorms in the vicinity
   c. Fronts (change in wind direction)
   d. Virga rain (temperature inversion), e.g., rain that evaporates before it reaches the ground, absorbs latent heat out of the surrounding air that creates denser/heavier air with an associated higher pressure causing a downdraft windshear.
   e. Land/sea breezes at coastal airports, especially at dawn and dusk periods
   f. Terrain, i.e., trees, mountains, or even surface obstructions

Note: Pilots are required to report windshear whenever they encounter it.

2. Aerodrome equipment/reporting. Many of the major aerodromes around the world have windshear measuring equipment.
3. Aircraft warning equipment. Modern aircraft have windshear warning systems, which usually are incorporated into the ground proximity warning system (GPWS).

How does windshear affect an aircraft?

Windshear is a change of wind speed and/or direction (including vertical updrafts/downdrafts) that affects the lift capability of an aircraft. The two main aerodynamic principles of this are as follows:

1. The aircraft's dynamic speed is reduced as a result of the reduction in wind speed.
2. The effective aircraft weight is increased as a result of experiencing a downdraft, which requires the pilot to recover the flight path by pitching the aircraft up. This is accomplished by deflecting the stabilizer/elevator upward (to get the aircraft to pitch up). However, deflecting the elevator/stabilizer upward creates a downward air

force on the control surface, which also has the effect of increasing the aircraft's effective weight.

These two properties both subject the aircraft to turbulence and affect the aircraft's lift potential, which brings the aircraft closer to its stall. (*See Q: How would you fly an approach if you suspect windshear? page 319.*)

A change in wind speed and/or direction (windshear) is in general terms manifested as flight path overshoot and undershoot effects. The actual windshear overshoot or undershoot effect depends on the following:

1. The nature of the windshear
2. Whether the airplane is climbing or descending through that particular windshear in which direction the aircraft is proceeding

An overshoot windshear effect may result from flying into

1. An increasing headwind
2. A decreasing tailwind
3. An updraft

An undershoot windshear effect may result from flying into

1. A decreasing headwind
2. An increasing tailwind
3. A downdraft

What is a microburst?

A microburst is a severe downdraft, i.e., vertical wind, emanating from the base of a cumulonimbus cloud during a thunderstorm.

Note: A microburst is a severe form of windshear.

Where do you find microbursts?

Microbursts are found close to, normally underneath, mature cumulonimbus clouds and are associated with thunderstorms.

The microburst downdraft is highly concentrated, typically only 5 km across, and often centered in the middle of a thunderstorm, surrounded on all sides by strong updrafts.

What do you know about microbursts?

(*See Q: What is a microburst? page 256.*) Microbursts are severe vertical downdrafts associated with mature cumulonimbus clouds. (*See Q: Where do you find microbursts? page 256.*) They are usually highly concentrated, only about 5 km across, and are found in the middle of a thunderstorm under a mature cumulonimbus cloud.

The mechanics that produce a microburst are found at the height of a mature cumulonimbus cloud's cycle; e.g., they only last for a few minutes (up to 10 minutes) because they are centered in the mature stage of a cumulonimbus cloud (where updrafts are produced that fuel the downdrafts), which itself only last for about 40 minutes. A mature cumulonimbus cloud produces strong continuous updrafts around its outer edges (and sometimes outside the cloud itself) that build up to fuel the downdrafts in the center of the cloud. The microburst is a result of the downdrafts breaking out of the base of the cloud being colder than the surrounding air because it has only been warmed at the saturated adiabatic lapse rate (SALR) during the descent within the cloud. Therefore, since it is colder, it is also denser/heavier than the surrounding air, and thus it continues down to the surface, where it often can be felt as the first gust. (*See Q: Describe the formation of a thunderstorm, page 232.*)

A microburst can be associated with precipitation or dry downdrafts. When they are at their most severe, microburst downdrafts can reach up to 3000 ft/min and have a wind speed of approximately 100 knots. Therefore, it is always best to wait for it to pass because the downdrafts and the rapid reversals of wind direction of a microburst can be fatal to an aircraft. [*See Qs: What identifies (are the indications of) a microburst? page 257; How does a microburst affect an aircraft? page 258.*]

What identifies (are the indications of) a microburst?

The typical indications of a microburst are

1. Mature cumulonimbus clouds with thunderstorm activity, especially rain
2. Roll cloud formation around a cumulonimbus cloud

Note: Roll clouds are formed by the turbulent outflow, associated with microburst downdrafts, from underneath a cumulonimbus cloud.

3. Virga rain, especially beneath or near to a cumulonimbus cloud

Note: Rain that evaporates before it reaches the ground absorbs latent heat out of the surrounding air and creates denser air with an associated higher pressure, causing the downdrafts to fall to the ground at an even greater rate. Therefore, virga rain indicates a possibly severe microburst.

4. Flight path and indicated airspeed (IAS) fluctuations, especially on the approach path (*See Q: How does a microburst affect an aircraft? page 258.*)
5. Wind direction and speed changes, which can be either reported or measured by aircraft systems, if fitted

How does a microburst affect an aircraft?

With a microburst out of the base of a cumulonimbus clouds, the downburst will spread out as it nears the ground and then rebound off the ground as an updraft. If an aircraft's flight path (approach) is beneath such a cloud, then the aircraft will experience the following characteristics:

1. Initially, when entering a zone underneath a cumulonimbus cloud with a microburst present, an aircraft will experience an updraft (as a result of the downburst spreading outward and rebounding off the ground as an updraft). This has the following effects on an aircraft's flight path:
   a. Energy gain from an increasing headwind will cause the aircraft's nose to rise.

Note: In an extreme case, the updraft may be severe enough to cause the aircraft to exceed its stalling angle.

   b. Airspeed (IAS) will rise.
   c. Rate of descent will be reduced.

   The overall effect in this portion of flight is to cause an overshoot effect as the aircraft tends to fly above the desired flight path.

Note: When such an updraft is encountered at the start of an approach underneath a cumulonimbus cloud, a pilot should be aware of the potential for the initial overshoot effect of the aircraft to be reversed as the aircraft proceeds along the flight path. This is known as *windshear reversal effect,* i.e., an overshoot followed by an undershoot.

2. Approaching the central area underneath a cumulonimbus cloud with a microburst present, an aircraft would start to experience a strong downdraft, which would be a severe change, over a short distance, having previously experienced an updraft. This has the following effects on an aircraft's flight path:
   a. There will be an energy loss from a reducing headwind, which will cause the aircraft's nose to fall.
   b. Airspeed (IAS) starts to fall (due to the reducing headwind).
   c. Rate of descent starts to increase, with a tendency to go below the glide path.

3. As the aircraft's flight path progresses to the far side of the cloud with a microburst, the aircraft will then experience a severe downdraft and a tailwind directly from the cloud base. This has the following effects on the aircraft's flight path:
   a. There will be energy loss from an increasing tailwind.
   b. Airspeed (IAS) continues to fall sharply.
   c. Rate of descent continues to increase with a tendency to go further below the glide path, with even the possibility of ground contact if not checked by initiating a missed approach, and even then, success depends on the power height and speed reserves available being sufficient enough to overcome the downdraft.

The overall effect in this portion of flight is to cause an undershoot effect as the aircraft tends to fly below the desired flight path.

It should be understood clearly that microbursts can be of such a strength that they can down an aircraft no matter how much power, time, and height the aircraft has available. Therefore, always, always avoid possible microburst areas. (*See Q: How would you fly an approach if you suspect a microburst? page 320.*)

What is the definition of wind gust factors?

Wind gust factors are an indication of how much wind speed change you can expect in varying wind conditions and are calculated by taking the difference between the maximum and minimum wind speeds in the gusts and dividing by the mean wind speed. Thus a wind gusting from 15 to 25 knots with a mean of 20 knots will have a gust factor of 10/20, or 0.5.

What do lenticular clouds indicate?

Lenticular (lens-shaped) clouds indicate standing (mountain) wave clear air turbulence (CAT). They are found at height in rising air above the downwind side of a range of hills, often extending for up to 100 nautical miles downward of a line of hills and at a height of up to 25,000 ft. [*See Qs: What is CAT, and give examples? page 259; Describe the formation of mountain (lenticular) clouds, page 230.*]

What is CAT, and give examples?

CAT is an acronym for clear air turbulence, i.e., no signs of visible moisture content in turbulent air. Examples of clear air turbulence are as follows:

1. *Low-level CAT.* This is caused by
   a. Temperature inversions

b. A difference between the surface and gradient wind due to the turbulent mixing of the boundary layer near the surface

c. Local surface winds, i.e., land/sea breezes

d. Terrain-generated winds (e.g., mountain/standing waves on the downwind side of a range of hills)

2. *Jetstreams.* The most severe CAT can be found level or just above the jet core in the warm air but on the cold air/polar side of the jet. (*See Qs: What is a jetstream? page 237; Where do you find jetstreams? page 237.*)
3. *Fronts.* An active front produces a marked horizontal windshear, which can produce CAT.
4. *Thunderstorms.* Serious CAT windshear can be found at low levels under and near thunderstorms, e.g., microbursts. (*See Q: What is a microburst? page 256.*)
5. *Wake turbulence.* (*See Q: What is wake turbulence? page 260.*)

Can you detect CAT?

Clear air turbulence (CAT) is one of the hardest forms of turbulence to detect. In fact, no specific equipment has been developed to detect CAT. Therefore, CAT can only be detected by using meteorologic appraisal of the prevailing conditions, typically shown on met charts, and identifying the locations of conditions that give rise to CAT.

Because CAT is a form of windshear, any reported wind and/or temperature change between two places and the presence of certain weather conditions, such as cumulonimbus clouds, lenticular clouds, fronts, or jetstreams, are good indications of possible CAT.

Note: In particular, a large, rapid fluctuation in the total air temperature (approximately ±10°C) in a few seconds is a very good indication of CAT, as are broken engine trails of a preceding aircraft.

What is wake turbulence?

Wake turbulence is the phenomenon of disturbed airflow; i.e., wing-tip vortices, created behind an aircraft's wing as the aircraft moves forward. (*See Q: What causes/are wing-tip vortices? page 10.*)

Note: Wake turbulence can be a serious hazard to lighter aircraft following heavier aircraft.

Wake vortex turbulence is present behind every aircraft, including helicopters, when in forward flight. It is extremely hazardous to a lighter aircraft with a smaller wingspan during takeoff, initial climb, final approach (including the circuit), and landing phases of flight following a heavier aircraft with a larger wingspan.

Note: Wake turbulence also can be encountered in the cruise but is not really a hazard, just an uncomfortable disturbance.

The characteristics of the wake vortex turbulence generated by an aircraft in flight are determined initially by the aircraft's gross weight, wingspan, airspeed, configuration, and attitude. That is, a heavy aircraft with a large wingspan creates the greatest wake turbulence (wing-tip vortices) because it creates the greatest pressure differential between the upper and lower surfaces of the wing.

The wake turbulence created from trailing wing-tip vortices can be strong enough to upset a following aircraft, especially a lighter aircraft, if it flies into them. To avoid wake turbulence accidents and incidents, air traffic control (ATC) delays the operation of lighter aircraft behind heavier aircraft for up to 5 minutes to allow the vortices that create the wake turbulence to dissipate.

Note: Aircraft of the same size or heavier aircraft following lighter aircraft do not suffer from dangerous levels of wake turbulence and therefore are not usually spaced as per International Civil Aviation Organization (ICAO) separation minima. (*See Qs: What are the ICAO final approach separation minima? page 289; How do you avoid wake turbulence? page 290.*)

Icing

What is icing?

The formation of ice (icing) is the change of the state of water to a solid form when the temperature is less than the freezing point of water, i.e., 0°C.

Note: There are three states of water: liquid (water drops), gas (water vapor), and solid (ice). Ice can form from either of the other two states of water:

1. Water vapor, by sublimation, causing hoar frost. (*See Q: What is sublimation? page 261.*)
2. Water droplets, by freezing rain or by freezing supercooled water drops (SWDs).

Note: Airframe icing occurs in freezing clouds when SWDs are present, causing rime and/or clear ice. [*See Q: What are supercooled water droplets (SWDs)? page 262.*]

What is sublimation?

Sublimation is the process of turning water vapor immediately into ice when the dewpoint/actual temperature is less than 0°C. The usual result of sublimation is hoar frost, which is in fact regarded as a type of icing.

Note: The dewpoint is now called the *frost point.*

What are supercooled water droplets (SWDs)?

Supercooled water droplets (SWDs) are liquid water drops sized between 40 and 2000 μm that exist in a nonfrozen liquid form in the atmosphere at temperatures down to as low as −45°C, well below the normal freezing point of water (0°C). This is known as being *supercooled.*

Such supercooled water drops will freeze only partially on contact with a colder, subzero surface, i.e., the skin of an aircraft or a propeller blade. Usually, a SWD will turn progressively into ice as it is washed back along the colder aircraft surface. This is so because the SWDs release latent heat (into the remaining water) as it starts to freeze (*see Q: What is latent heat? page 222*), which reduces the rate of freezing of the remaining water/liquid, allowing it to stay in a liquid state for longer, which in turn gets spread backwards by the airflow along the aircraft's surface. As the remaining liquid/water comes into direct contact with another part of the subzero surface, it in turn freezes progressively. This can cause long trails of ice to build up on and from the leading edges, which the deicing systems of the aircraft must be capable of removing. For every one degree of super cooling, 1/80 of the water drop will change to ice on impact. Obviously, the larger the supercooled water drops, the more severe is the icing.

Note: SWDs do not exist below approximately −45°C, where any moisture is already an ice crystal in the atmosphere, and do not stick to the aircraft's airframe. Therefore, airframe icing is only possible between 0 and −45°C.

What conditions present an icing risk?

Icing from water drops that are present in clouds and as rain pose a risk to aircraft when the total air temperature (in flight) or outside air temperature (on the ground) is between +10°C and −40/45°C.

What hazards to aviation does icing cause?

Ice buildup on an aircraft or within its engine induction system can be a serious hazard to flight safety, particularly at slow flight phases of flight, e.g., takeoffs and landings, because of the following effects:

1. *Adverse aerodynamic performance effects.* Ice buildup on an aircraft's wings seriously disrupts the airflow pattern, causing it to separate at a significantly lower angle of attack than it would from a clean wing. This results in
   a. Reduced stalling angle
   b. Reduced maximum lift capability
   c. Increased stall speed
   d. Increased drag

   e. Reduced lift (loss of lift) at a given angle attack, especially at higher angles of attack
2. *Control surface effects.* Ice formation may freeze on control surfaces, leading to restricted movement and even loss of control in extreme circumstances.
3. *Increase in aircraft weight effects.* The buildup of ice on an aircraft will increase the overall weight of the aircraft, with all the associated effects of higher weights, i.e., higher stall speed, extra lift required, etc. In addition, this uncontrolled increase in weight may change
   a. The position of the aircraft's center of gravity
   b. The balance of the various control surfaces and propellers, causing severe vibration and/or control difficulties
4. *Reduced engine power.* Ice buildup in the jet engine intake or piston engine carburetor can restrict the airflow into the engine, causing a power loss and even engine failure in extreme circumstances. (*See Q: What is carburetor icing? page 264.*)
5. *Vent blockage effects.* Ice buildup on pitot and static probes will produce errors in the aircraft's pressure-driven flight instruments, airspeed indicator (ASI), altimeter, and vertical speed indicator (VSI). (*See Qs: What are the ASI, altimeter, and VSI indications and actions for a blocked pitot and/or static probe? pages 129, 131, and 133.*)
6. *Degraded navigation and radio communication effects.* If ice builds up on navigation and radio aerials, then these systems will be degraded and unreliable.

An ice-laden aircraft may even be incapable of flight. If the ice or frost buildup on the leading edges and upper surfaces of the wings is too great, then the lift produced is reduced and insufficient to balance the increased aircraft weight. In addition, if the ice builds up on the engine intake, the power output may be degraded to the extent that it is insufficient to balance the increased drag from the ice on the airframe.

What is a deicing system?

(*See page 174.*)

What is an anti-icing system?

(*See page 175.*)

In what conditions would you expect icing, and when should you turn on the engine anti-icing?

(*See page 324.*)

What is carburetor icing?

Ice formation can occur in the engine induction system and in the carburetor of piston engines, particularly in the venturi and around the throttle valve, where acceleration of the air can produce a temperature fall by as much as 25°C. This, combined with the heat absorbed as the fuel evaporates, can cause serious icing, even when there is no visible moisture present. Such a buildup of ice in a carburetor can disturb or even prevent the flow of air and fuel into the engine manifold, causing it to lose power, run roughly, and even to stop the engine in extreme circumstances. This effect is called *carburetor icing.*

Throttle icing (i.e., around the throttle valve) is more likely to occur at low power settings (e.g., descents), when the partially closed butterfly creates its own venturi cooling effect.

When would you expect carburetor icing in a piston engine?

(*See page 325.*)

What actions should you take to prevent or remove carburetor/throttle icing?

(*See page 325.*)

Visibility

What is atmospheric/meteorologic visibility?

Meteorological visibility is defined as the greatest horizontal distance at which a specified object can be seen in daylight conditions. It is therefore a measure of how transparent the atmosphere is to the human eye.

What conditions reduce visibility?

Visibility is reduced whenever particles are present in the atmosphere that absorb the light, e.g., water, ice, pollution, sand, dust, volcanic ash, etc. Poor visibility is usually associated with stable air, where the moisture and/or atmospheric contamination is kept in situ, especially at low levels. Unstable air produces a convection current that carries moisture and contamination aloft and thus produces good visibility at low levels.

Visibility may be further reduced by the position of the sun. Visibility may be much greater flying "downsun," where the pilot can see the sunlit side of objects, than when flying into the sun. As well as reducing the visibility, flying into the sun may cause a glare. If you are landing into the sun, consideration should be given to altering the time of arrival.

What is the most important visibility to the pilot?

The most important visibility to the pilot is from the aircraft to the ground, i.e., the slant or oblique visibility, especially during and in the direction of takeoff and landing. This may be very different from horizontal visibility distance (normal meteorologic measured visibility direction). For example, low-level stratus, smog, or fog may severely reduce the slant visibility (especially on an approach) when the vertical visibility might be unlimited.

How is visibility reported?

Visibility is reported in the following ways:

1. General visibility, from overall meteorologic reports
2. Runway visual range (RVR) for instrument landings

   General visibility. In general meteorologic reports, the visibility passed is always the *least* distance visible from the point of the observation in all directions. Visibility is reported in kilometers or, in very poor conditions, in meters.

   Runway visual range (RVR). (*See Q: What is RVR? page 278.*)

What is RVR?

(*See Chapter 9, "Flight Operations and Technique," page 278.*)

How is RVR reported?

(*See Chapter 9, "Flight Operations and Technique," page 279.*)

For fog questions, see subchapter, "Moisture and Clouds," page 226.

What is smog, and how is it formed?

Smog is a combination of smoke (and/or other airborne particles) and fog. (*See Q: What is fog? page 231.*)

Smog is usually found under an inversion layer, which acts like a blanket, stopping vertical convective currents. (*See Q: What is a temperature inversion/layer? page 225.*)

Climatology

Climatology questions can be about either (1) General weather mechanics (which are covered throughout Chapter 8, "Meteorology and Weather Recognition," or (2) weather conditions in specific parts of the world. Obviously, it is impossible to list the weather phenomena throughout the year for every airport location around the world in this section. Therefore, it is left to the good sense of the reader to identify the locations and their weather conditions that he or she is likely to be asked about.

What is the thermal equator?

The thermal equator is the position of the maximum thermal temperature around the earth's surface.

Note: It should not be confused with the global equator, which is the greatest parallel of latitude distance around the earth's surface.

The position of the thermal equator moves according to the sun's heating, which varies widely with season. In the summer season, the thermal equator moves toward the pole (in both the northern and southern hemispheres) due to the greater heating experienced at higher latitudes.

In addition, the seasonal temperature range over the land is greater than over the sea. Therefore, the thermal equator varies considerably compared with the global equator over the land but remains close to the global equator over the sea. (*See Q: What is the ITCZ? page 266.*)

What is the ITCZ?

The intertropical convergence zone (ITCZ) is where converging air masses meet near the thermal equator. Like the thermal equator, ITCZ movement is a function of seasonal heating that is much greater over the land than over the sea. Over South America and southern Africa, ITCZ movement is large, especially in the summer season, whereas over the Atlantic Ocean, its movement is small. In other words, it is stable. (*See Q: What is the thermal equator? page 266.*) The effect of the ITCZ determines the weather pattern over a significant portion of the globe.

What are tropical revolving storms?

Tropical revolving storms (TRSs) are deep, intense depressions, i.e., lows, found in equatorial regions around the intertropical convergence zone (ITCZ). (*See Qs: What is a pressure system? page 243; Describe the characteristics of a low-pressure system, page 243.*) They are known as *cyclones* in the Indian and Pacific Oceans, *hurricanes* in the Caribbean and Americas, and *typhoons* in the China Sea area.

Tropical revolving storms start at the edge of the ITCZ (*see Q: What is the ITCZ? page 266*) as it retreats in late summer or early autumn.

This is so because they are enormous heat engines, and they take their energy from water vapor off the warm sea, which is condensed aloft, releasing latent heat. Sea temperatures of at least 27°C are needed to form a TRS, which is a temperature that the sea achieves only after prolonged heating, i.e., after a long, hot summer. For this reason, TRSs do not form over cold seas and die out when they pass over land or move to colder sea region. However, TRSs do not form at the equator because the coriolis effect is zero at the equator, so they form at approximately 5 to 20 degrees of latitude, where the sea temperature is high enough and the coriolis effect is present. [*See Q: What is the coriolis force (or geostrophic force)? page 236.*] On crossing land, TRSs can cause immense damage.

Weather Forecasts and Reports

What is the primary method of preflight meteorologic briefings for aircrew?

The primary method of preflight meteorologic briefing for aircrew in most parts of the aviation world is self-briefing. This is done in one of two ways:

1. Using facilities, information, and documentation routinely available or displayed in aerodrome briefing areas.

Note: Computerized systems providing specific route meteorologic conditions are commonplace in most of the major airports around the world.

2. Using a telephone service to call an aviation authority met office; e.g., area forecasts (TAFS) or reports (METARS) are available directly from a meteorologic officer, and/or area forecasts are available from an airmet telephone recording system, especially in the United Kingdom. (*See Q: What is an AIRMET? page 267.*)

What is an AIRMET?

An AIRMET is a recorded telephone message that gives the meteorologic forecast for a particular area. It can be accessed by area, and telephone numbers for different areas are listed on an AIRMET Areas Chart.

What is a meteorologic report?

A meteorologic report is an observation of the actual weather at a specific time, i.e., either past or present.

What are the common aviation meteorologic reports?

The common aviation meteorologic reports are

1. METARS, SIGMETS, and SPECI (*See Qs: What are METARS? page 268; What is a SIGMET? page 269; What are SPECIs? page 269.*)
2. Automatic Terminal Information Service (ATIS)
3. In-flight weather reports (i.e., Volmets, ATIS, and by radio communications with an air traffic service unit or flight information service)

What are METARs?

A metar is a written, coded routine aviation weather report for an aerodrome. It is an observation of the actual weather given by a meteorologic observer at the aerodrome.

Note: Cloud base in a metar is given above ground level (AGL).

Decode the following METAR: METAR: EGCC/0920Z/210/15 G27/1000 SW/R24/P1500M/SHRA/BKN 025 CB/08/06/Q1013 or 29.92/RE TS/WS TK OF RWY 24/NOSIG.

EGCC	Location identifier (Manchester, U.K.)
0920Z	Time the report was taken (09.20 hours UTC)
210/15 G27	Wind direction and speed (210 degrees true at 15 knots, gusting to 27)
1000 SW	Horizontal visibility (1000 m to the southwest)
R24/P1500m	Runway visual range (runway 24 plus 1500 m of visibility)
SHRA	Weather (rain showers)
BKN 025 CB	Clouds (broken at 2500 ft with cumulonimbus clouds)
08/06	Temperature/dewpoint (8°C temperature, 6°C dewpoint)
Q1013 or 29.92	QNH (1013 millibars or 29.92 in)
RE TS	Recent weather (recent thunderstorms)
WS TK OF RWY 24	Windshear (windshear report on takeoff runway 24)
NOSIG	Trend (no significant change)

What does *trend* mean in a meteorologic report?

A weather trend is usually attached to an aerodrome weather report, i.e., METAR (*see Q: What are METARs? page 268*), and is commonly referred to as a *landing forecast.*

The trend is a forecast of any significant weather changes expected in the next 2 hours after the time of the report and is described using the normal coded weather format and abbreviations. If no significant change is expected, the observation (report) will be followed by NOSIG (no significant change) as the trend.

Because a trend forecast period is much shorter than a normal aerodrome forecast; i.e., a TAF, it should be much more accurate.

Note: Cloud bases in a trend are given above aerodrome level (AAL).

A trend can only be given by a qualified forecaster, whereas a report can be given by just an observer.

What is a SIGMET?

A SIGMET is a meteorologic report that advises of significant meteorologic (SIG/MET) conditions that may affect the safety of flight operations in a general geographic area, i.e., en route or at an aerodrome.

The criteria for raising a SIGMET include

1. Active thunderstorms
2. Tropical revolving storms
3. Severe line squalls
4. Heavy rain
5. Severe turbulence
6. Severe airframe icing
7. Marked mountain waves
8. Widespread dust or sandstorms

What are SPECIs?

A SPECI is an Aviation Selected Special Weather Report for an aerodrome. It is generated whenever a critical meteorologic condition exists, e.g., windshear, microbursts, etc. It is similar in presentation to a METAR.

What in-flight weather reports can you access?

In-flight weather reports that can be accessed include

1. Flight information service or air traffic (control) service
2. ATIS (*See Q: What is ATIS? page 269.*)
3. VOLMET (*See Q: What is VOLMET? page 270.*)

What is ATIS?

The Automatic Terminal Information Service (ATIS) is a prerecorded tape broadcast on an appropriate VOR or VHF channel to reduce the workload on air traffic control (ATC) communications frequencies that give current information on aerodrome operations and weather.

Note: Some of the larger aerodromes have both an arrival and a departure ATIS.

The ATIS message is changed with any significant change in the reported conditions, and each new message has a new alphabetical designator prefix, e.g., alpha, beta, etc., to distinguish the current from the old.

What is VOLMET?

VOLMET is a continuous broadcast on a VHF/HF frequency that includes

1. The actual weather report
2. The landing forecast
3. A forecast trend for the 2 hours following
4. A SIGMET (significant weather, if any) of several selected aerodromes that produce meteorologic reports within a given region

What is a meteorologic forecast?

A forecast is a prediction, or prognosis, of what the weather is likely to be for a given route, area, or aerodrome.

What are the common types of aviation forecasts?

The common types of aviation forecasts are

1. Area forecasts for preflight briefings (Note that cloud bases in area forecasts are given above mean sea level.)
2. Aerodrome forecasts (e.g., TAFS and trends) (*See Qs: What is a TAF? page 270; What does trend mean in a meteorologic report? page 268.*)
3. Special forecasts (*See Q: What is a special forecast, and when is it requested? page 271.*)

What is a TAF?

A Terminal Aerodrome Forecast (TAF) is a coded routine weather forecast for an aerodrome. It is a weather forecast given by a qualified meteorologic forecaster based at the aerodrome.

Note: Cloud base in a TAF is given above aerodrome level (AAL).

TAFs are usually issued for a 9-hour period and updated every 3 hours. However, they may be issued for up to 24 hours with updates every 6 hours, but their accuracy is not as high.

Decode the following TAF: TAF: EGCC/15 0600/0716/210/15/5000/SHRA/BKN 025/TEMPO/1116/3000/SHRA/PROB30/1416/TSRA/BKN 006CB.

EGCC	Location identifier (Manchester, U.K.)
15 0600	Time the report was taken (15th day, 0600 hours UTC)
0716	Validity time of forecast (0700 to 1600 hours UTC)
210/15	Wind direction and speed (210 degrees true at 15 knots)
5000	Horizontal visibility (5000 m)
SH RA	Weather (rain showers)
BKN 025	Clouds (broken at 2500 ft)
TEMPO	Variation (*See Q: What is the definition of Tempo? page 271.*)
1116	Validity time of TEMPO (1100 to 1600 hours UTC)
SH RA	Weather (rain showers)
PROB30	Probability (*See Q: What is the definition of PROB? page 272.*)
1416	Validity time of PROB (1400 to 1600 hours UTC)
TS RA	Weather (thunderstorm, rain showers)
BKN 006CB	Clouds (broken at 600 ft with cumulonimbus clouds)

What is a special forecast, and when is it requested?

Special forecasts are meteorologic forecasts for flights over long routes outside the coverage of the local countries' area forecast.

Note: Special forecasts also can include forecasts (TAFS) for the departure, destination, and up to three alternative aerodromes.

What is the definition of BECMG in a forecast?

BECMG means "Becoming" and is followed by a four-figure time group, which is two different whole UTC hours. It indicates a permanent change in the forecasted conditions occurring at some time during the specified period.

What is the definition of TEMPO in a forecast?

TEMPO relates to a temporary variation in the general forecasted weather lasting less than 1 hour or, if recurring, lasting in total less than half the trend or TAF period it is included within.

Note: A TEMPO can relate to improvements as well as deteriorations in wind, visibility, weather, or clouds.

Once the TEMPO weather events have finished, the original prevailing weather reasserts itself.

What is the definition of INTER in a forecast?

INTER relates to intermittent variations in the general forecasted weather that are more frequent than TEMPOs, i.e., conditions fluctuating almost constantly.

Note: An INTER can relate to improvements as well as deteriorations in wind, visibility, weather, or clouds.

Once the INTER weather events are finished, the original prevailing weather reasserts itself.

What is the definition of GRADU in a forecast?

GRADU relates to gradual change in the original weather at approximately a constant rate throughout the period or during a specified part thereof to a different and new weather condition.

What is the definition of RAPID in a forecast?

RAPID relates to a rapid change, in less than half an hour (30 minutes), of the original weather to a new and different prevailing weather condition.

What is the definition of PROB in a forecast?

PROB is used in weather forecasts when the forecaster is uncertain if the weather conditions will occur or not, and therefore, he or she assesses the *probability* of them occurring as less than 50 percent. (If greater than 50 percent, then it would be listed as a TEMPO.) For example, PROB30 means a 30 percent chance of the weather conditions occurring.

What is the definition of CAVOK?

When the following conditions occur simultaneously:

1. Visibility equal or greater than 10 km
2. No clouds below 5000 ft or below the highest minimum sector safe altitude (MSA), whichever is the greater, and no cumulonimbus clouds at any altitude
3. No precipitation, thunderstorms, shallow fog, or low drifting snow

then the term *CAVOK* is used to replace visibility, RVR, weather, and clouds in meteorologic reports and forecasts.

Note: CAVOK does not mean clear blue skies.

Chapter

9

Flight Operations and Technique

Flight Operations

What are NOTAMs, and how are they distributed?

NOTAMs are NOTices to AirMen. They contain information on any aeronautical facilities, services, procedures, or hazards and timely knowledge required by people concerned with flight operations. They may be given as either

1. A class 1 NOTAM, distributed by teleprinter for urgent matters
2. A class 2 NOTAM, distributed through the post for less urgent matters

What are AICs?

Aeronautical Information Circulars (AICs) are published monthly and concern administrative matters and advance warnings of operational changes. They draw attention to and advise on matters of operational importance.

When must you file a flight plan?

You must file a flight plan for flights

1. Within controlled airspace notified as instrument flight rules (IFR) only, irrespective of whether instrument meteorologic conditions (IMC) or visual meteorologic conditions (VMC) exist
2. Within other controlled airspace in IMC or at night, excluding special visual flight rules (SVFR).
3. Within certain special rules airspace, irrespective of weather conditions
4. Within upper information regions (UIRs), i.e., above FL245

5. When the destination is more than a certain distance and the maximum total weight authorized exceeds 5700 kg
6. When you cross-flight information regions (FIR) boundaries in most parts of the world
7. When you intend to use the air traffic advisory service on an advisory route (ADR)

What is ETOPS?

Extended twin operations.

An operator is granted permission to operate a twin-engined aircraft type on flights in which the aircraft is more than 60 minutes away from a suitable alternative aerodrome in the event that the aircraft suffers an engine failure en route. (*See Q: What is an adequate/suitable aerodrome, especially with regard to ETOPS diversion/alternate aerodromes? page 274.*)

ETOPS approval of up to 180 minutes allows twin engined aircraft to cross most of the Pacific and Atlantic Oceans.

Note: Recently, some airlines have applied for 207 and 240 minute of ETOPS to allow better trans-Pacific routings.

What is an adequate/suitable aerodrome, especially with regard to ETOPS diversion/alternate aerodromes?

An adequate/suitable aerodrome for ETOPS diversions is one in which

1. Aircraft performance is suitable for the airfield
2. Adequate emergency facilities are available at the aerodrome
3. Adequate aerodrome lighting facilities are available for night flights
4. A basic instrument approach is available for any expected instrument meteorologic conditions (IMC)
5. The aerodrome is open

What are the various ETOPS categories?

Normal extended twin operations (ETOPS) categories, vary between 60 and 180 minutes. However, in recent years (from 1999 onward), applications have been made for up to 207 and 240 minute of ETOPS by trans-Pacific airlines.

What is OCH(A)?

OCH is the *obstacle clearance height* above the aerodrome level (and so has relevance when using QFE). OCA is the *obstacle clearance altitude* above mean sea level (and so has relevance when using QNH).

What is the minimum-height/altitude rule?

In order to comply with the instrument flight rules (IFR) both inside and outside controlled airspace, and without prejudice to the usual low-flying rules, the minimum height rule dictates that an aircraft should not fly at a height of less than a 1000 ft above the highest obstacle within a distance of 5 nautical miles of the aircraft unless

1. It is necessary for the aircraft to do so in order to land
2. The aircraft is flying on a route notified (NOTAM, AIP) for the purpose of this rule (this may include controlled airspace such as terminal control areas and airways)
3. The aircraft has been otherwise authorized by the competent authority
4. The aircraft is flying at an altitude not exceeding 3000 ft above mean sea level (MSL) and remains clear of clouds and in sight of the surface

An additional International Civil Aviation Organization (ICAO) height rule increases this height to 2000 ft above the highest point in mountainous terrain.

What lighting designations are on air navigation (i.e., man-made) obstacles?

Obstacles greater than 492 ft (150 m) are lit by high-intensity flashing white lights by day and night. Any failed lights are NOTAMed. Obstacles less than 492 ft (150 m) but higher than 300 ft, are lit when the obstacle is considered significant with medium-intensity flashing red lights at night. These lights are *not* NOTAMed when they fail.

What is MSA?

Minimum sector altitudes (MSAs) are published on

1. *Instrument approach charts.* They provide at least 300 m (1000 ft) vertical clearance within 25 nautical miles of the homing facility for the particular instrument approach. If the aircraft remains at or above the relevant MSA, then it should remain clear of terrain and obstacles as it tracks toward the aerodrome prior to commencing the approach. Some aerodromes have MSAs that apply to all sectors; however, most aerodromes have different MSAs for different sectors depending on the direction from which the aircraft is arriving and the terrain over which it must cross.
2. *En route charts.* These show grid moras (or minimum sector safe altitudes). When determining your current MSA or grid mora altitude from an en route chart, you should take the most restrictive MSA of all the adjacent grid moras to your present position.

What is MEA?

Minimum en route altitude (MEA) is the safe altitude within the airway, i.e., 5 nautical miles either side of the airway centerline, and a minimum altitude at which radio reception is guaranteed.

What are the IFR flight levels?

The instrument flight rule (IFR) flight levels are based on the quadrantal and semicircle rule whenever an aircraft is more than 3000 ft above mean sea level or above the appropriate transition level, whichever is the higher, and the aircraft is in level flight. The following levels flown are based on the aircraft's *magnetic track:*

Flights at levels below 24,500 ft (quadrantal rule):

Magnetic Track, degrees	Cruising Level, ft
0–to 89	Odd thousand
90–179	Odd thousand plus 500
180–269	Even thousand
270–359	Even thousand plus 500

Therefore, vertical separation is only 500 ft. However, usually only thousand-feet levels are used.

Flights at levels above 24,500 ft (semicircular rule):

Magnetic Track, degrees	Cruising Level	RVSM Levels	
0–179	250		
	270		
	290	290	310
	330	330	350
	370	370	390
	410	410	430
180–359	260		
	280		300
	310	320	340
	350	360	380
	390	400	420

Therefore, vertical separation is only 1000 ft up to FL290 and 2000 ft above FL290 in non-RVSM airspace and only 1000 ft in RVSM airspace at all levels.

What pressure settings are flight levels based on?

Flight levels are based on the standard altimeter setting of 29.92 inches of mercury (inHg) or 1013.2 millibars (hPa).

Why are flight level intervals increased to 2000 ft above FL290 in non-RVSM airspace?

Vertical separation is increased to 2000 ft above FL290 in non-RVSM airspace because of increased altimeter errors due to the lower air density experienced at these higher levels.

Note: Reduced vertical separation minimum (RSVM) to 1000 ft separation above FL290 is granted to aircraft with advanced and more accurate altimeters, especially on crowded routes, e.g., trans-Atlantic routes.

What is the lowest usable flight level (FL)?

The lowest usable en route flight level must be at least 500 ft above the absolute minimum altitude.

Note: Minimum altitude on an airway is at least 1000 ft above the highest obstacle within 15 nautical miles of the airway centerline (hence you comply with the minimum height rule). (*See Q: What is the minimum height rule? page 275.*)

Therefore, terrain/obstacle clearance is at least 1500 ft at the lowest usable flight level.

What are the two quantities known as weather minima?

1. Decision height (DH) or minimum decision altitude (MDA) (*See Q: What is decision height/minimum decision altitude? page 277.*)
2. Runway visual range (RVR) or visibility (*See Qs: How is visibility reported? page 265; What is RVR? page 278.*)

What is decision height (DH)/minimum decision altitude (MDA)?

DH is the wheel height above the runway elevation at which a go-around must be initiated by a pilot unless adequate visual reference has been established and the position and approach path of the aircraft have been assessed visually as satisfactory to safely continue the approach and landing.

DH is the height above the ground; i.e., it is measured off the radio altimeter or with the local QFE pressure setting off the barometric altimeter.

MDA is the altitude measured using the local QNH pressure setting; i.e., it is the height above sea level, or MDA = airport elevation + height above the ground.

How is a decision height (DH) or minimum decision altitude (MDA) calculated for a precision approach?

[*See Q: What is decision height (DH)/minimum decision altitude (MDA)? page 277.*] DH minimum for an instrument-rated precision approach is calculated as follows:

Step 1. Take the higher of (a) obstacle clearance height (OCH) for the aid and aircraft category [*see Q: What is OCH(A)? page 274*] or (b) precision approach system minimum, i.e., ILS (CAT 1), 200 ft; PAR, 200 ft; and MLS, 200 ft.

Step 2. Then add 50 ft for altimeter position error correction (PEC), especially for light aircraft.

Note: Many operators of advanced aircraft do not add 50 ft for PEC because their altimeter systems are extremely accurate and therefore do not suffer from PEC.

How is a minimum decision altitude (MDA) calculated for a nonprecision approach?

[*See Q: What is decision height (DH)/minimum decision altitude (MDA)? page 277.*] MDA minimum for an instrument-rated nonprecision approach is calculated as follows: Take the higher of (a) obstacle clearance height (OCH) for the aid and aircraft category [*See Q: What is OCH(A)? page 274*] or (b) nonprecision approach system minimum, i.e., localizer only, 250 ft; SRA terminates at ½ nautical mile, 250 ft; SRA terminates at 1 nautical mile, 300 ft; SRA terminates at 2 nautical miles, 350 ft; VOR, 300 ft; and NDB, 300 ft.

Note: Many operators add a further 50 ft, to the MDA, especially for large aircraft.

How is visibility reported?

(*See Chapter 8, "Meteorology and Weather Recognition," page 265.*)

What is RVR?

Runway visual range (RVR) is a highly accurate instrument-derived visibility measurement that represents the range at which the runway's high-intensity lights can be seen in the direction of landing along the runway. Its readings are transmitted to the air traffic controller, who informs the pilot. RVR is used, when available, as a visibility minimum for low-visibility precision approach landings in preference to the general visibility measurement, which also may be reported.

RVR values are measured at three points along the runway:

1. Touchdown point
2. Midpoint
3. Endpoint

How is RVR reported?

Runway visual range (RVR) is reported at up to three points on the runway:

1. At the touchdown zone
2. At the midpoint
3. At the stop end

whenever it is detected as being less than 1500 m.

Midpoint and stop-end values are only reported if they are less than the touchdown zone and less than 800 m or if they are less 400 m. When all three values are given, the names are omitted. RVR is measured in steps of 25 m up to 200 m, in steps of 50 m up to 800 m, and thereafter in steps of 100 m.

Note: The runway designator may follow the quoted RVR; e.g., RVR 300/25 means RVR 300 m on runway 25. (*See Q: What RVR limits are required for LVP instrument approaches and takeoffs? pages 289 and 299.*)

Give the definitions of radar control, radar advisory, and radar information services.

1. Radar control service is available wherever radar coverage exists in controlled airspace, i.e., airways, terminal control zones, aerodrome traffic zones, and control areas, whereby air traffic control (ATC) is responsible for
   a. Monitoring and separation from other aircraft
   b. Radar vectoring
   c. Controlled airspace crossing
   d. Navigation assistance
   e. Weather information

f. Hazard warnings
g. Emergency assistance

Note: Instructions from ATC under a radar control service have to be adhered to. However, it remains the responsibility of the aircraft commander, *not* ATC, to maintain terrain clearance even when under radar control.

2. Radar advisory service. The radar controller will use radio communications to provide
   a Traffic information
   b. Advisory avoiding action necessary to maintain separation from other aircraft
3. Radar information service. The radar controller will use radio communications to provide traffic information only (no avoiding action will be offered).

What is radar vectoring, and what is required for radar vectoring to be carried out?

Radar vectoring occurs when a radar controller passes to an aircraft a heading to steer; e.g., Delta 204 steer heading two seven zero.

Note: Bear in mind that the radar controller is trying to get you to achieve a particular track over the ground. However, because the controller does not know precisely what the wind drift is, he or she occasionally will request a modification to your heading to achieve and/or maintain the desired track.

What is the standard circuit direction?

Left-hand direction.

Note: However, some runways might have a nonstandard right-hand circuit pattern. This is typical for aerodromes with neighboring noise-sensitive areas, high ground, or other restrictive airspace that precludes the use of a left-hand circuit.

What does HST mean on an airfield runway chart?

High-speed turn off a runway.

What is a rate 1, 2, and 3 turn?

Rate 1 is a 3 degree per second or 180 degree per minute turn. Rate 2 is a 6 degree per second or 360 degree per minute turn. Rate 3 is a 9 degree per second or 540 degree per minute turn.

What is the altitude effect on wind direction and speed?

The wind normally backs in direction (veers in direction in the southern hemisphere) and decreases in speed during a descent as you get near to the ground. (*See Q: Describe the characteristics of a surface wind, page 238.*)

What is the night effect of wind direction and speed?

The wind at night normally backs further in direction (veers in direction in the southern hemisphere) and decreases more in speed as you get nearer to the ground during a descent. (*See Q: Describe the diurnal variation of the surface wind, page 239.*)

What does it mean if you have a port wind in the northern hemisphere?

This means that you are flying toward a low-pressure system, resulting in a descending flight path for a constant altimeter pressure setting, e.g., 1013 millibars/29.92 inHg.

Why is the correct rotation rate important, especially on large jet aircraft?

On modern aircraft, the importance of the correct rotation rate at the correct speed (V_r) ensures that the aircraft leaves the ground within the correct distance (runway performance) and achieves the correct initial climb speed (V_2).

Note: The correct pitch attitude after lift off also has to be achieved to maintain the V_2 speed. There is a natural rotation rate appropriate for each aircraft type. However, under high-performance conditions (low weight, low altitude, and low temperature, etc.), the aircraft will have a high rate of acceleration, and the rotation rate will need to be that much faster. However, under these conditions, it is easy to overrotate and tail strike the aircraft, so beware. Similarly, under a low-performance condition (maximum weight, high altitude, and high temperature, etc.), a lower rate of rotation will be required, reflecting the slower acceleration of the aircraft.

What causes the noise from a jet aircraft?

The noise from a jet aircraft is caused by the shear effect at the boundary of the jet efflux. The higher the power, the faster is the speed of the jet efflux, the greater is the shear effect, and the greater is the noise.

Note: The development of bypass engines has reduced the speed of the jet efflux and therefore the noise considerably. (*See Q: What is the principle of a bypass engine? page 65.*)

Describe a typical noise abatement technique.

There are two main parameters affecting noise, which in practice can be handled to achieve a noise abatement profile:

1. Noise is proportional to the power being developed by the engines.
2. Noise is inversely proportional to the distance between the noise source and the listener.

Therefore, to reduce noise, we must reduce power and get as far away from noise-sensitive areas as quickly as possible. The following basic technique is the only one that could be used if the airfield is completely surrounded by noise-sensitive areas. After leaving the runway, the technique is split into two separate parts.

1. The noise abatement first segment adopts a steep climb at full power to gain as much height as possible before the listening posts short of the noise-sensitive areas.

Then the flight path changes to

2. The noise abatement second segment, where the engines are throttled back to climb power while the aircraft maintains its initial high attitude. This profile continues until either a declared height has been reached or the noise-sensitive area has been cleared, when a lower climb attitude is adopted.

This leads into the departure profile's third phase, acceleration and cleanup, followed by the fourth phase, en route climb. (*See Q: Describe departure profile segments 1 to 4, page 204.*)

Why on a short sector would you climb to FL330?

To gain a better specific fuel consumption (SFC), which improves the higher the altitude. Additionally, you gain a higher true airspeed (TAS) but not necessarily a higher ground speed. (*See Q: What advantages does a jet engine gain from flying at a high altitude? page 71.*)

What is the glide distance for an aircraft at 30,000 ft?

Determine the best glide speed (minimum drag speed) and its rate of descent; altitude divided by rate of descent (ROD), e.g., 30,000/1000 ft/min = 30 minutes; Time × speed per minute, e.g., 30 × 5 nautical miles/min = 150 nautical miles.

Note: Weight does not affect the best glide distance. (*See Q: What is the effect of weight on its glide range? page 2.*)

Why does an aircraft descend quicker when it is lighter?

Because an aircraft is restricted to a maximum speed during a descent, the heavier aircraft has to maintain a lower rate of descent than a lighter aircraft; otherwise, it would overspeed. Remember, heavier aircraft have a greater momentum, and this weight-driven momentum will produce a greater speed in a vertical dive. Therefore, a heavier aircraft has to start its descent earlier than a lighter aircraft because it has to maintain a shallower descent. (*See Q: How does weight affect an aircraft's flight profile descent point? page 18.*) In other words, a lighter aircraft can descend later and quicker than a heavier aircraft because it can maintain a greater vertical descent profile without overspeeding.

Why does an aircraft have to descend earlier when it is heavier?

(*See Qs: How does weight affect an aircraft's flight profile descent point? page 18; Why does an aircraft descend quicker when it is lighter? page 283.*)

What visual clues on landing should you look for?

The main visual clue during a landing is the pilot's judgment of a 3-degree glide slope; i.e., a flat-looking runway means that you are low; a steep-looking runway means that you are high. This judgment only comes with experience. Initial judgment of an appropriate glide slope may be facilitated by aids such as VASIs and PAPIs or by positioning the aircraft at predetermined heights above known ground features or distance from touchdown points.

During the final approach stages, the extended sides of the runway intersect at the horizon, and texture gradients in the surrounding terrain also indicate horizon location. However, these cues are only accurate when the terrain and runway are level. Sloping runways and terrain may produce incorrect estimates of horizon location by the pilot and result in an inaccurate judgment of the approach slope.

To prevent the final approach angle from varying, the pilot should aim at his or her projected impact point because this is the point on the ground away from which visual texture flows. As long as the visual texture flows away from this point, and as long as the visual angle between this point and the horizon remains constant, the approach is progressing normally.

How do you calculate headwind, tailwind, and crosswind component?

To calculate headwinds, tailwinds, and crosswinds, you should use a chart, such as Fig. 9.1, and enter the tower-reported wind; i.e., enter at the relative angle of the wind to the aircraft nose and/or runway, and follow along that line until you reach the speed for the reported wind. Then read the headwind or tailwind off the side of the chart and the crosswind off the bottom of the chart.

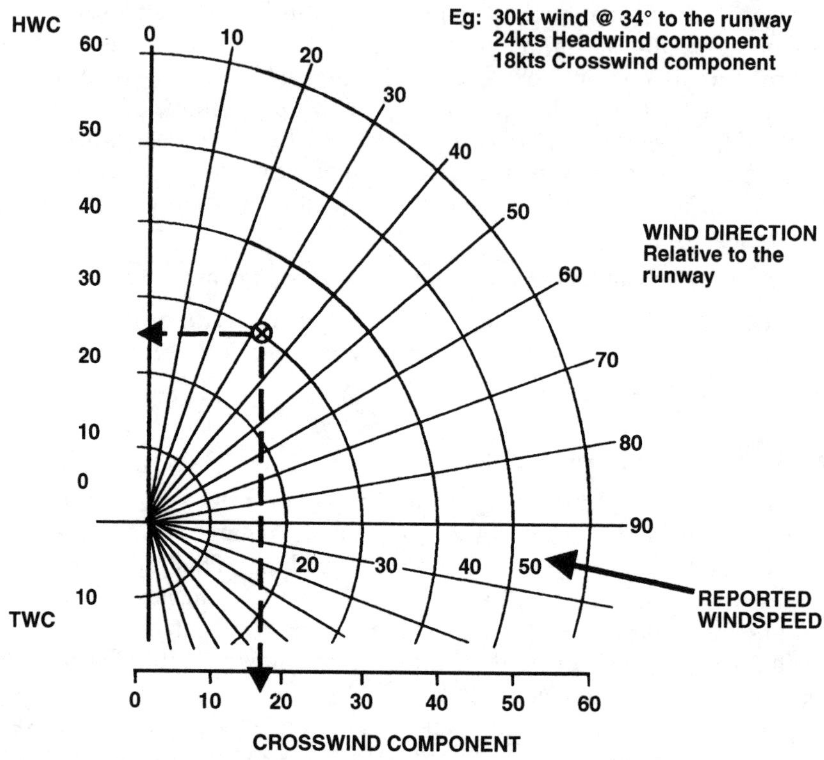

Figure 9.1 Wind component.

How do you fly a crosswind approach and landing, yaw, or wing down?

Whatever the company policy is.

Which is the more difficult landing, a left or right crosswind?

A crosswind from the left of the aircraft's nose in the northern hemisphere is more difficult for the following reason: The wind backs in direction during a descent in the northern hemisphere. Therefore, a crosswind from the left of the aircraft's nose becomes greater nearer the ground because it backs in direction from (*A*) to (*B*) in Fig. 9.2 resulting in a lower headwind component and a greater crosswind component.

In contrast, a right-sided crosswind component will become farther from the ground because it backs from (*C*) to (*D*), resulting in a greater headwind component.

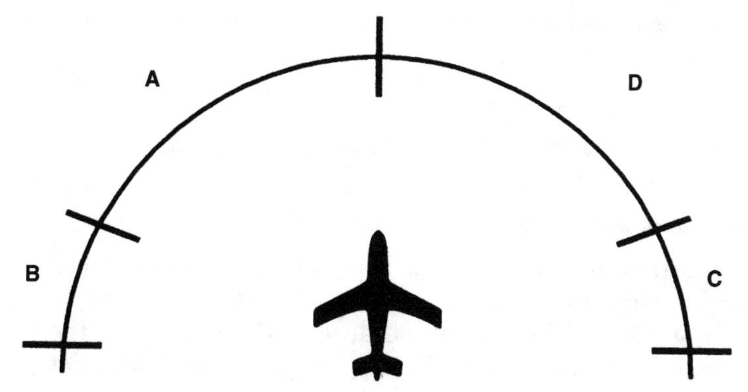

Figure 9.2 Variation of crosswind direction in a descent.

In the southern hemisphere, a crosswind from the right creates the more difficult crosswind landing because the wind veers in direction during a descent in the southern hemisphere.

How do you correct for a high sink rate on approach?

If you need extra lift to correct for the unique problem of a high sink rate on approach, you can create the extra lift required in one of two ways:

1. *By increasing incidence.* In this case, you must then increase thrust to counter the extra drag from the higher incidence, or the resulting sink rate will be higher. This is true except when you are trading an excess of airspeed for height, but then be careful to monitor your speed in case it decreases below your approach speed.
2. *By increasing airspeed.* This is obtained by increasing thrust while keeping all other parameters substantially constant. A heavy aircraft takes a lot of accelerating, so if this option is exercised, a lot of thrust will be needed.

Note: If you see the need for more thrust for any reason, then apply enough of it early enough, especially for jet engines. This is so because of the poor response from a low rpm setting and the comparatively small amounts of thrust produced for a given power lever movement in the lower ranges of power settings.

What are the wind gust corrections applied to the approach speed?

½ × stable wind (minimum of 5 knots) + full gust value up to a predetermined, type-specific limit of approximately 20 to 25 knots. For example, basic approach speed of 120 knots, headwind component 20 knots gusting to 28 knots:

$$120 + 10\ (\tfrac{1}{2}\ \text{stable wind}) + 8\ (\text{gust value}) = 138\ \text{knots (correct approach speed)}$$

Why is it important not to lose speed on the approach?

It is important not to lose speed on the approach, especially in jet swept-wing aircraft, because the approach speed is only just above minimum drag speed (VIMD), and because of the relative flatness of the drag curve, especially around VIMD, the jet aircraft does not produce any noticeable changes in flying qualities. Speed is unstable below VIMD, where an increase in thrust has a greater drag penalty for speed gained, with a net result of losing speed for a given increase in thrust. Therefore, if you lose speed below your approach speed, you could fall into this situation, which results in a decreasing speed and altitude.

Note: VIMD is a higher speed on a jet aircraft because the swept wing is more efficient against profile drag, and therefore, the minimum drag speed is typically a higher value. (*See Q: Describe the drag curve on a jet aircraft, page 8.*)

Recovery from this situation is further compounded because of the jet engine's slow response from idle, and if you have slipped up the back of the drag curve, you need a lot of thrust to counter its effect, and you need it quickly. This all makes a recovery rather marginal when close to the ground, and therefore, it is doubly important not to lose speed on the approach.

For a propeller-driven aircraft with a well-defined VIMD point on its drag curve (which is due to the marked increase in profile and induced drag either side of VIMD), any deviation of airspeed below VIMD on the approach is very well defined by the steep increase in the drag curve. (*See Q: Describe the drag curve on a propeller-driven aircraft, page 7.*)

In addition, the slipstream effect from the propeller also makes recovery from a falling airspeed on the approach much better. Therefore, although it is always important to maintain a stable speed on the approach, it is not as marginal on a propeller-driven aircraft as it is on a jet.

What is ground effect, and how is it caused?

Ground effect is the cushioning of the aircraft on the air between it and the ground when the aircraft is close to the ground, such as in the flare to land.

Note: In effect, the cushioning effect produces more lift.

Ground effect is caused by rising warm air off the ground, by a reduction in the amount of downwash behind the wings, and by the tendency of the aircraft not to slow (decreased drag) resulting from the reduction of wing-tip vortices.

How does ground effect influence landing distance?

Ground effect will, if allowed to, cause the aircraft to float along the runway past its touchdown point. This results in an increased landing distance. (*See Q: What is ground effect, and how is it caused, page 286.*)

What is the most efficient system for stopping at high speed?

Reverse thrust. (*See Q: How does an auto brake system decelerate an aircraft? page 168.*)

Heavy jet aircraft have a high kinetic energy (momentum) during a landing roll or an aborted takeoff, and the most efficient method of stopping that maintains the initial deceleration rate when the aircraft is at a high speed is best achieved by reverse thrust for two reasons:

1. The net amount of reverse thrust increases with speed because the acceleration imposed on the constant mass flow is greater. This is so because the aircraft's forward speed is additional when using reverse thrust as opposed to subtracting when in forward thrust.
2. The power produced is greater at high speeds because of the increased rate of work done. This means that the kinetic energy of the aircraft is being destroyed at a greater rate at higher speeds.

Why should you not use reverse thrust at low speeds?

The reason to cancel reverse thrust before you reach low speeds, i.e., approximately below 70 knots, is for the following engine-handling considerations:

1. On a four-engined or greater aircraft, the reversed flow of the inner engines, being ahead of the outer engines, tend to upset the intake flow of the outer engines if reverse thrust is left engaged at low speeds.
2. Any engine in isolation will tend to breathe its own reversed exhaust at low speeds, which may blow surface contamination into the engine.

In addition, if reverse thrust is maintained at a low speed on a contaminated surface, e.g., snow, dust, etc., then forward visibility is compromised because a cloud is blown forward by the reverse thrust.

Can you use reverse thrust in flight?

No. Reverse thrust is not certified for in-flight deployment on most aircraft types, except for a very few exceptions, which use it as an air brake.

What selection will give you TOGA?

Takeoff/go-around (TOGA) selection depends on the aircraft type. The two common systems that will give TOGA thrust and flight director pitch commands are

1. Moving the thrust levers into a forward TOGA gate, as on the Airbus 320.
2. Pressing the TOGA switches on the thrust levers, as on the Boeing 737.

What is a typical engine fire drill?

A typical engine fire drill is as follows:

Thrust lever	Close
Auto throttle (if engaged)	Disengage
Start lever/switch	Off or cut out
Engine fire warning switch	Engage

If the fire or overheat warning light remains illuminated:

Extinguisher number 1	Discharge

If the fire or overheat warning light still remains illuminated:

Extinguisher number 2	Discharge

If the fire or overheat warning light still remains illuminated: *land at the nearest airport.*

Note: Various type-specific system paper checks would follow.
For a multicrew cockpit, the drills, both memory and paper checks, would be carried out on a challenge and response basis.

What are the International Civil Aviation Organization (ICAO) aircraft category weight definitions?

Heavy (H)	136,000 kg
Medium (M)	136,000 kg and 7000 kg
Light (L)	7000 kg or less

Note: These weights refer to an aircraft's maximum takeoff weight.

Aircraft greater than 136,000 kg, i.e., heavy, are required to be announced as heavy on initial contact with an air traffic control (ATC) unit.

Why should you maintain the minimum approach separation?

To avoid the wake vortex/turbulence from the preceding aircraft.

Note: Wake turbulence can be a serious hazard to lighter aircraft following heavier aircraft.

What are the International Civil Aviation Organization (ICAO) final approach separation minima?

The ICAO final approach spacing minima are as follows:

		Separation	
Leading Aircraft	Following Aircraft	Nautical Miles	Minutes
Heavy	Heavy	4	—
Heavy	Medium	5	2
Heavy	Light	6	3
Medium	Heavy	3	—
Medium	Medium	3	—
Medium	Light	5	3
Light	Heavy	3	—
Light	Medium	3	—
Light	Light	3	—

Note: These minima may be applied when an aircraft is operating directly behind another aircraft and when crossing behind at the same altitude or less than 1000 ft below whenever a controller believes there is a potential hazard due to wake vortex/turbulence. They may even be increased at the discretion of the controller or at the request of the pilot. Such requests must be made before entering a runway or commencing a final approach.

Many of the very busy commercial airfields may space aircraft closer than these separation limits; therefore, you need to be aware of the increased possibility of encountering wake vortex/turbulence.

The separation minima stated cannot entirely remove the possibility of a wake turbulence encounter. The objectives of the minima are to *reduce* the probability of a vortex wake encounter to an acceptable low level and to minimize the magnitude of the upset if an encounter does occur.

Care should be taken when following any substantially heavier aircraft, especially in conditions of light winds. *The majority of serious incidents occur close to the ground, especially when the winds are light.*

What are the International Civil Aviation Organization (ICAO) minimum departure separation criteria?

The ICAO departure spacing minima are as follows:

Leading Aircraft	Following Aircraft Departing from the Same Position	Minimum Separation at the Time Aircraft Are Airborne
Heavy	Medium (small in UK) or light	2 minutes
Medium (or small in UK)	Light	2 minutes

Leading Aircraft Departed Using the Full Length of the Runway	Following Aircraft Intermediate Position on the Runway	Minimum Separation at the Time Aircraft Are Airborne
Heavy	Medium (small in UK) or light	3 minutes
Medium (or small in UK)	Light	3 minutes

How do you avoid wake turbulence?

The main aim of wake turbulence avoidance is to avoid passing through it at all. There are two methods of avoiding wake turbulence:

1. Separation minima (time)
2. Alteration to the flight path

(*See Q: What is wake turbulence? page 260.*)

1. *Separation minima.* These are a time and distance technique employed by air traffic control (ATC) to allow the wake turbulence of preceding aircraft to dissipate. This technique typically is used during final approach and takeoff phases of flight, where the flight path track cannot be altered.

Note: Wake turbulence can be a serious hazard to lighter aircraft following heavier aircraft. (*See Q: What are the ICAO final approach/departure separation minima? pages 289 and 290.*)

2. *Alteration to the flight path.* This is the ultimate action to be taken to avoid wake turbulence, and this is the pilot's responsibility. Therefore, you must understand and be able to visualize the formation and movement of the invisible wing-tip vortices from preceding aircraft to be able to avoid these danger areas, especially heavier aircraft at low speeds and high angles of attack at low altitudes, which produce the greatest wake turbulence. Wake vortex generation begins when the nose wheel lifts off the ground on takeoff and continues until the nose wheel touches down on landing. Vortices will tend to lose height slowly, up to 1000 ft, and will drift with the wind, which is especially important in crosswind conditions because you can encounter wake turbulence when passing downwind of a preceding aircraft.

Note: A *crosswind* will cause the vortices to drift downwind of the preceding aircraft. A *headwind* or *tailwind* will carry the vortices in the direction of the wind. A *nil, light,* or *variable* wind condition will make the vortices just hang around. This condition can be very dangerous, and it is worth considering delaying your takeoff or approach or even changing runway.

Instrument Flight Rules (Instrument Procedures and Flight Technique)

Instrument flight rule (IFR) theory questions are likely to be asked of pilots looking for their first-ever position. This is so because an airline employer has to teach you to fly its aircraft type and operating procedures, not IFR.

If you are currently flying, then the following questions probably will be somewhat basic and insulting to your professional standing. However, if you have been out of regular flying for some time, then this section might be a worthwhile review.

Many of the answers in this section are given as possible answers. This is so because there can be several different techniques (which are all correct) that can accomplish the procedure relating to the question. The answers are offered to the reader as suggestions only, and if the suggested technique is different from the reader's own training and knowledge of a particular procedure, then the reader would be wise to keep with the technique with which he or she is conversant.

What are SIDs?

Standard instrument departures (SIDs). A SID details a specific initial route or track from a particular aerodrome runway, often with altitude and, occasionally, speed constraints at specific points along the track.

What are STARs?

Standard instrument arrivals (STARs). A STAR details a specific final route or track onto a particular runway approach, often with altitude and, occasionally, speed constraints at specific points along the track.

What is a holding procedure?

A holding procedure is a predetermined maneuver that keeps an aircraft within a specified airspace while awaiting further clearance. A holding procedure/pattern generally is a racetrack shape.

What is the standard holding pattern direction?

The standard holding pattern direction is right-hand turns.

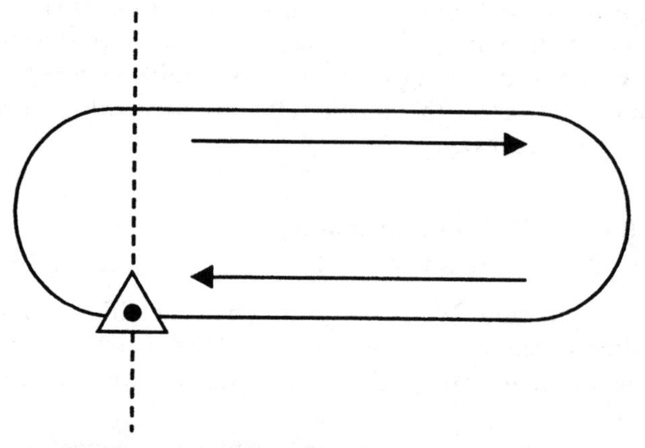

Figure 9.3 Standard holding pattern direction.

Therefore, the nonstandard holding pattern direction is left-hand turns.

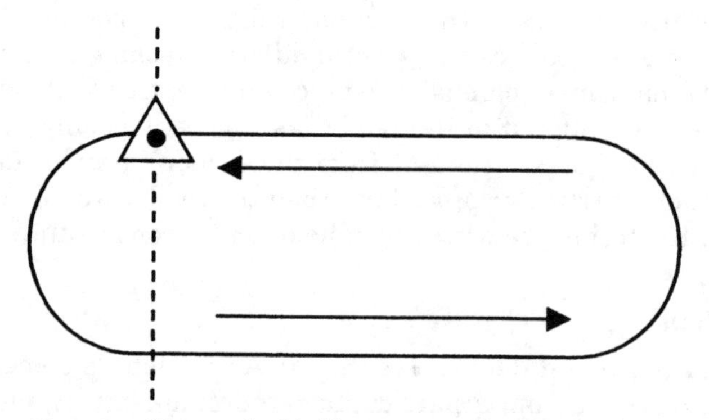

Figure 9.4 Nonstandard holding pattern direction.

What are the three entry procedures into a holding pattern?

The three entry procedures into a holding pattern are based on the sector of entry. The three sector regions have been devised based on the direction of the inbound holding track and an imaginary line angled at 70 degrees to the inbound holding track through the fix.

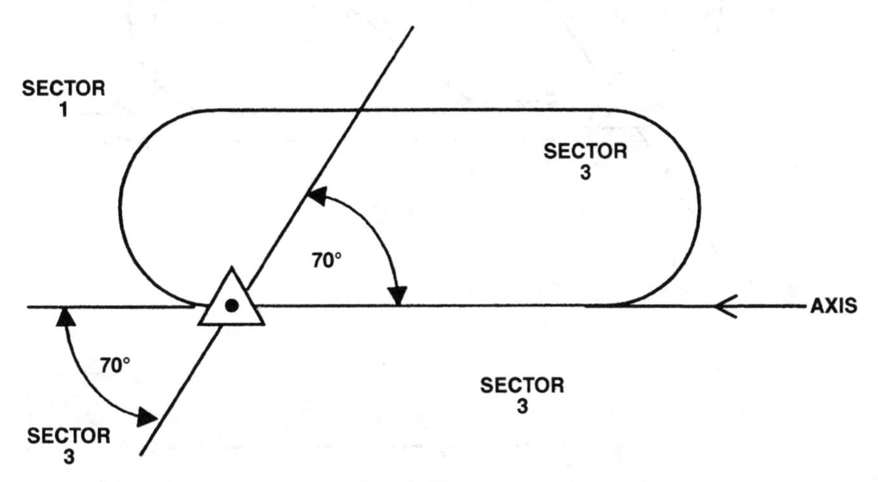

Figure 9.5 The three entry sectors for a holding pattern.

However, although the sectors are determined by track lines from the fix, the sector from which the aircraft joins the holding pattern, and therefore the entry procedure employed, is determined by the aircraft's *heading* to the fix.

Note: Be aware that you may be approaching a holding fix with a track in one sector but a heading in another sector. It may feel strange, but the sector that relates to the aircraft's heading determines the entry procedure employed.

Sector 1 procedure: Parallel entry. Fly to the fix, and turn onto an outbound heading to fly parallel to the inbound track on the non-holding side for a period of 1 minute plus or minus 1 second per knot wind correction. Then turn in the direction of the holding side through more than 180 degrees to intercept the inbound track to the fix. On reaching the fix, turn to follow the holding pattern.

Sector 2 procedure: Teardrop entry. Fly to the fix, and turn onto a heading to fly a track on the holding side at 30 degrees offset to the reciprocal of the inbound track for a period of 1 minute plus or minus 1 second per knot wind correction. Then turn in the direction of the holding pattern to intercept the inbound track to the fix. On reaching the fix, turn and follow the holding pattern.

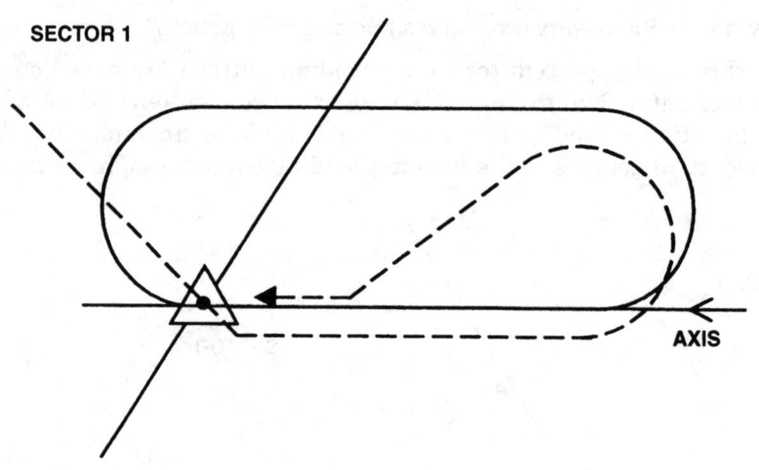

Figure 9.6 The sector 1 parallel entry.

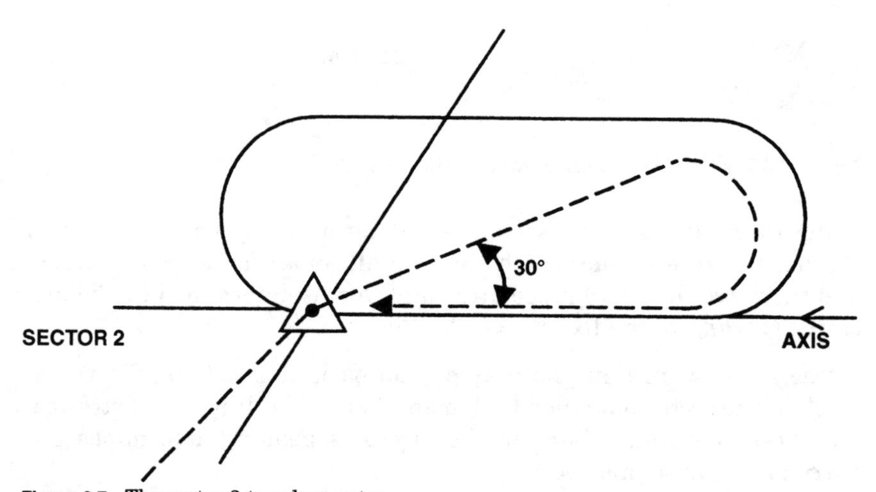

Figure 9.7 The sector 2 teardrop entry.

Sector 3 procedure: Direct entry. Fly to the fix, and turn to follow the holding pattern. On the face of it, the sector 3 direct entry procedure is the easiest to carry out. But a little thought will show that when joining from the extremities of the sector area, it is necessary to apply some finesse to the procedure. If a full 180-degree or greater turn is required over the fix when joining to take up the outbound heading, then commence turning immediately when you are overhead of the fix.

If, however, the turn onto the outbound heading is less than 180 degrees but greater than 70 degrees, then hold your heading for an

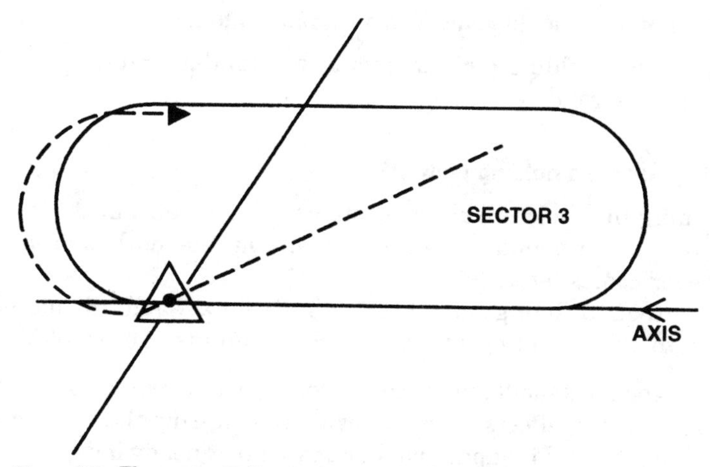

Figure 9.8 The sector 3 direct entry, >180-degree turn.

appropriate time past the fix, approximately 5 to 15 seconds, before commencing a rate 1 turn onto an outbound track. For example, for a turn of 170 degrees, hold your heading for 5 seconds; for a turn of 70 degrees, hold your heading for 15 seconds. (*See Figure 9.9*)

If, however, the turn onto the outbound heading is close to 180 degrees, then you are already very close to or on the inbound track of the holding pattern. That is, your joining track is close to the hold's inbound track. Therefore, simply turn onto a normal outbound track from over the fix, with only small finesse adjustments, as if you are already in the holding pattern.

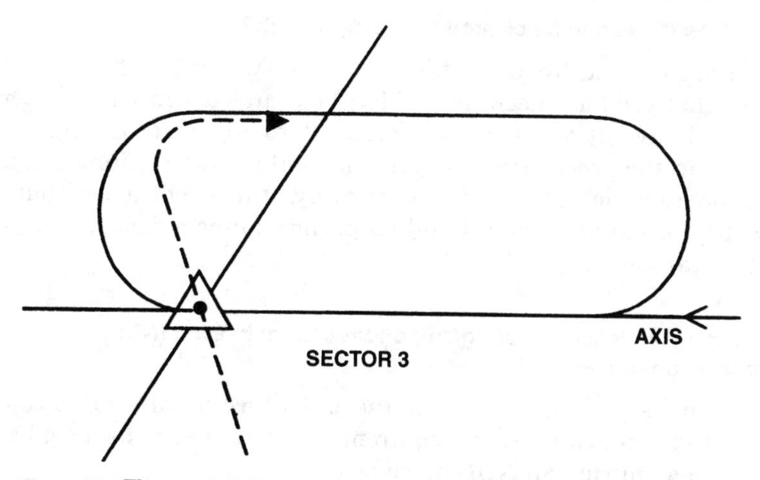

Figure 9.9 The sector 3 direct entry, <180-degree and >70-degree turn.

What rate of turn should you use in a holding pattern?

All turns in a holding pattern should be standard rate 1 turn, i.e., 3 degrees per second

How do you time a holding pattern?

The timing in a hold should be commenced from abeam the fix at the start of the outbound leg or on attaining the outbound heading, whichever comes later.

The outbound timing should be 1 minute for a standard holding pattern up to and including 14,000 ft and 1½ minutes above 14,000 ft.

Note: The pilot should make corrections to the timings (and headings) to correct for the effects of the known winds during the entry and the when flying the holding pattern. For tailwind or headwind components, a reasonable correction is to reduce the outbound leg by 1 second per knot of tailwind and to increase the outbound leg by 1 second per knot of headwind.

With distance-measuring equipment (DME) available, the outbound leg may be expressed in terms of distance. Where this is done, care should be taken that at least 30 seconds should be available on the inbound leg after completion of the inbound turn and that the slant range is taken into account.

Note: When a clearance is received specifying a departure time from the holding fix, then the pilot should adjust his or her pattern within the limits of the established holding procedure in order to leave the holding fix at the time specified.

What does it mean to be cleared for the approach?

If you are cleared by air traffic control (ATC) for an approach, this means that you have been cleared for the complete procedure in terms of lateral navigation as well as vertically to any altitude/height constraints in the procedure from your present altitude/height and then onto the approach descent to the runway. It does not mean that you have been cleared to land. Landing permission is a distinct and separate clearance.

When can you descend during an nondirectional beacon (NDB) nonprecision procedure?

After you have been cleared for the procedure by air traffic control (ATC), there are two further requirements that need to be meet before you descend during an NDB procedure:

1. You must be at the descent point as defined on the procedure. This may be expressed as a fix, distance, or time.
2. You must be within ±5 degrees of the automatic direction finder (ADF) final approach track.

These requirements relate to any point on the procedure, i.e., the outbound procedure, and not just the inbound track.

How is the MAP defined on an nondirectional beacon (NDB) approach?

The missed approach point (MAP) for a nonprecision approach is defined by a fix, either a facility, a distance from a fix, or a time from a fix, or the start of the inbound descent (MAPt). You cannot descend below the calculated minimum descent altitude (height) [MDA(H)] unless you become visual; however, you can level off and maintain the MDA and track into the MAP in the hope of becoming visual. If not visual at the MAP, a missed approach should be initiated.

Note: Most commercial operators would carry out a descent that mirrored a constant glide slope that reached the MDA at the MAP. Thus the missed approach is initiated at the MDA.

However, if a turn is specified in the missed approach procedure, then it should not be initiated until the aircraft has passed the MAP and is established in the climb.

On an instrument landing system (ILS) approach, when can you descend on the glide path?

You can descend on the glide path when

1. You have been cleared for the ILS procedure.
2. You have captured the localizer, within ±5°C.

What are the visual and aural indications of instrument landing system (ILS) marker beacons?

Outer marker	Blue light	Low-pitch dashes, the letter *O,* ———, in Morse code
Middle marker	Amber light	Medium-pitch dots and dashes, the letter *C,* -.-., in Morse code. (Note that the Morse letter *M,* ——, is not used because other than the pitch it would be indistinguishable from *O* when transmitted repeatedly.)
Inner marker	White light	High-pitch dots, the letter *I,* .. , in Morse code

How can you calculate the distance from the threshold at which you would intercept the glide slope?

If you divide the height by the glide slope angle times 100, you should get the distance from the threshold in nautical miles at which you should intercept the glide slope. For example:

$$3000 \text{ ft}/(3 \times 100) = 10 \text{ nautical miles (approximately)}$$

Likewise, if you wish to calculate the height at which you would intercept the glide slope at a particular distance from the threshold, multiply the distance in nautical miles by the glide slope angle times 100. For example:

$$10 \text{ nautical miles} \times (3 \times 100) = 3000 \text{ ft (approximately)}$$

How do you calculate glide slope angle if an approach is only given as a gradient?

Glide slope angle = gradient % × 0.57

How can you calculate the approximate rate of descent for a 3-degree glide slope?

5 × ground speed = ROD (feet per minute) required for a 3-degree glide slope. For example:

$$5 \times 140 = 700 \text{ ft per minute ROD}$$

What are the Cat I, II, and III ILS International Civil Aviation Organization (ICAO) approach limits?

CAT I	Decision height (DH) not lower than 200 ft	Runway visual range (RVR) at least 550 m at touchdown or visibility 800 m RVR/VIS at midpoint N/A
CAT II	DH 200 ft or lower but not less than 100 ft	RVR at least 300 m at touchdown RVR at least 150 m at midpoint
CAT IIIa	DH 100 ft or lower but not less than 50 ft	RVR at least 200 m at touchdown RVR at least 150 m at midpoint
CAT IIIb	DH less than 50 ft (down to 50 ft)	RVR at least 50 m at touchdown RVR at least 50 m at midpoint
CAT IIIc	DH 0 ft	RVR 0 m

Note: CAT IIIc has never been in commercial operation because it would require an automatic taxiing guidance system. CAT IIIb, however, is used widely.

Note: Company, local authorities, or specific airfield limits may be greater than these limits. If this is the case, then the higher limits take precedence.

What are the runway visual range (RVR) minima for a low-visibility takeoff?

With high-intensity runway center lights available, the RVR minimum for takeoff is 150 m RVR at the touchdown point (or 125 m under some authorities), and the midpoint RVR must be at least equal to the touchdown RVR. The reported stop-end RVR can be less than the takeoff minimum.

With *no* high-intensity runway center lights available, the RVR minimum for takeoff is 200 m RVR at the touchdown point, and the midpoint RVR must be at least equal to the touchdown RVR. The reported stop-end RVR can be less than the takeoff minimum.

Describe the requirements for a CAT II or III approach.

The requirements to carry out a CAT II or III approach in addition to the normal CAT I approach are as follows:

1. The aircraft's CAT II or III systems are certified and operational.
2. The runway's ground CAT II or III equipment is certified, operational, and protected (namely, low-visibility procedures in force).
3. The crew is qualified and current.
4. Weather minima are above the approach ban prior to commencing the final approach.

Note: If the reported runway visibility range (RVR) drops below the approach minimum during the final approach (i.e., past 1000 ft approach landing height), the approach can be continued.

5. Alternate airfield weather is above CAT I minimum.

What do you know about CAT II and III approaches?

CAT II and III approaches are different categories of precision instrument landing system (ILS) approaches. They are based on the principle that the short-range visual reference required (RVR) will be present at the decision height (DH). They are collectively referred to as *low-visibility procedures* (LVPs). (*See Qs: What are the Cat I, II, and III ILS ICAO approach limits? page 298; Describe the requirements for a CAT II or III approach, page 299.*)

CAT II and III approaches usually are terminated with the execution of an autoland. (*See Q: What is a fail passive / operational autopilot/landing system? page 162.*)

What is the final approach fix (FAF)?

The final approach fix (FAF), denotes the start of the final approach segment of either a precision or nonprecision approach. It is usually denoted as either a locator marker or a final approach distance.

Note: On some instrument precision approaches, the FAF is the same as the initial approach fix. For example, the initial approach fix (IAF) may be the locator outer marker when flying outbound, and the FAF is the same locator outer marker when flying inbound.

Where no FAF is shown, the final approach segment commences at the completion of the procedure turn and when the aircraft is within ±5 degrees of the final approach track, whereupon final descent may be commenced.

Note: The FAF always should be crossed at or above the specified intermediate height before final descent is initiated.

What is alert height?

Alert height is a specified radio height (typically approximately 100 ft) based on the characteristics of the aircraft and its fail operational landing system, above which, if a failure occurred in one of the required redundant operational systems (fail operational systems) during a CAT II or III approach, the approach would be discontinued, and a go-around would be executed.

The exception would be a reversion to a higher decision height, i.e., CAT I or II, if the aircraft were still above the approach to landing height, normally 1000 ft, whereby it was considered that enough time existed to safely transfer to a new approach limit.

Also, if a failure occurred in one of the required redundant operational systems below the alert height, the approach also would be continued, the considered reason being that the probability and effects of a failure are such that if it should occur, the operation could be continued safely with the remaining system, "a second failure below the artificial horizon (AH) being considered extremely improbable." Improbable but not impossible!

What is MABH?

Minimum approach breakoff height. That is, the lowest height of the wheels above the ground that if a go-around were initiated without external references in normal operation, the aircraft would not touch the ground during the procedure, or, with a critical engine failure, it can be demonstrated that an accident is improbable during a go-around.

Describe the SRA nonprecision approach procedure.

A surveillance radar approach (SRA) is a nonprecision approach used at certain aerodromes where a radar controller can provide tracking guidance and desired height information down a final approach.

Note: An SRA radar only provides the controller with returns showing the aircraft's lateral position, not its vertical position. (*See Q: What are the minima for an SRA procedure? page 301.*)

An SRA is carried out by the pilot under guidance from the radar controller, who passes the following on the VHF-COM radio:

1. Radar vectors to steer the aircraft onto a final approach
2. Advice to the pilot about when to start a (3-degree) descent
3. Horizontal tracking instructions in the form of turns to make and headings to steer to maintain the inbound track. (e.g., "Turn right 5 degrees, heading two six five")
4. Vertical navigation (descent) guidance in the form of range to touchdown and the desired height at that range (the heights given are for QFE set on the altimeter subscale, e.g., "Range 4 miles, height should be 1230 ft.)

An SRA is achieved by a ground radar unit being aligned with a specific runway. Therefore, it is only available on certain runways at certain aerodromes.

What are the minima for an SRA procedure?

The surveillance radar approach (SRA) system minima are

SRA terminating at ½ nautical mile = 250 ft MDH

SRA terminating at 1 nautical mile = 300 ft MDH

SRA terminating at 2 nautical miles = 350 ft MDH

Describe the PAR approach procedure.

A precision approach radar (PAR) approach is a precision approach. It is similar to a surveillance radar approach (SRA) (*see Q: Describe the SRA nonprecision approach procedure, page 301*) except that it has a different ground-based radar system that provides returns in vertical (slope) navigation as well as horizontal (azimuth) navigation.

Therefore, the controller can give voice instructions in both horizontal and vertical parameters.

This enables the PAR to be a much more accurate and controlled approach system, and as such, it is classified as a precision approach with a corresponding lower minimum decision altitude (MDA). (*See Q: What is the minimum for a PAR procedure? page 302.*)

What is the minimum for a PAR procedure?

The PAR system minimum is 200 ft MDH.

What is a DME arc procedure?

A distance-measuring equipment (DME) arc is based on a navigation aid, e.g., a nondirectional beacon (NDB) or VOR, and is used to intercept a final approach track. It is flown by maintaining two parameters:

1. A constant DME reading from the aid on which the arc is based.
2. By keeping the aid at a constant 90 degrees off the aircraft's heading.

What is a procedure turn, and when is it used?

There are two types of procedure turns:

1. The 45-degree procedure turn, which consists of
   a. An outbound track from the fix (usually the reciprocal of the approach track) to a set distance or timed point
   b. An outbound turn of 45 degrees away from the outbound track for either 1 minute (or 1 minute and 15 seconds) from the start of the turn or 45 seconds from wings level once the turn has been completed (plus or minus wind correction in terms of drift and time)

Note: Left or right in a description of the procedure turn refers to the direction of this initial turn.

   c. A 180-degree turn in the opposite direction to intercept the inbound track

2. The 80/260-degree procedure turn consists of an outbound track from the fix (usually the reciprocal of the approach track) to a set distance or timed point.
   a. An outbound turn of 80 degrees away from the outbound track
   b. Followed almost immediately by a 260-degree turn in the opposite direction to intercept the inbound track

Note: If the initial turn is into a strong headwind, then the heading can be held for a brief time before the 260-degree inbound turn is commenced. For a tailwind, stop the turn slightly before completing the 80-degree turn, and gently roll immediately onto the inbound turn.

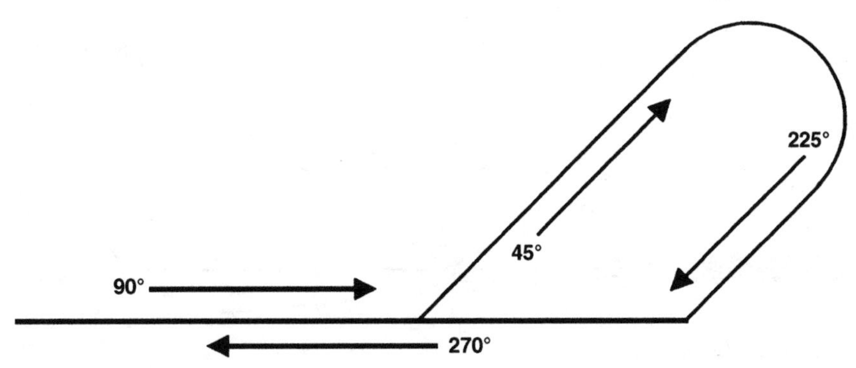

Figure 9.10 A 45-degree procedure turn.

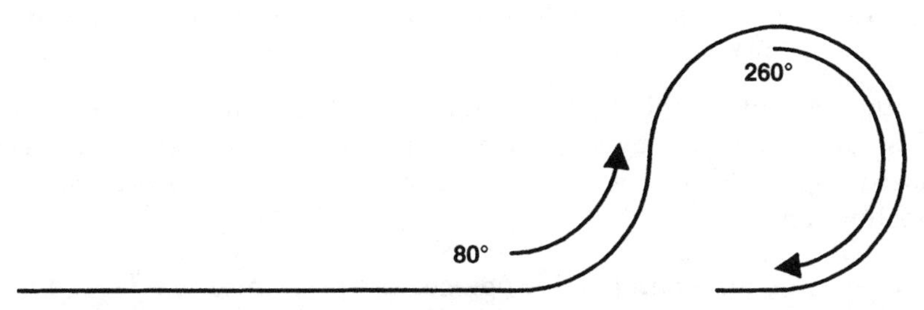

Figure 9.11 An 80/260-degree procedure turn.

Procedure turns are used when no suitable fixes permit a direct entry onto an instrument approach procedure.

What is a (teardrop pattern) base turn and when is it used?

(*See Figure 9.12*)
The teardrop base turn is used to reverse the direction by more than 180 degrees onto an inbound final approach track and consists of the following:

1. A specified outbound track from a fix to a set distance or time

Note: Different outbound tracks exist for different categories of aircraft based on their speed to allow sufficient space for the inbound turn (i.e., faster aircraft require a larger turning radius).

2. Followed by a turn to intercept the inbound track of an instrument approach.

Can you extend the outbound leg of a holding pattern as a base turn?

You can extend the outbound leg of a holding pattern as an alternative procedure when the inbound track of a hold is the same as the nondirectional beacon (NDB) or instrument landing system (ILS) approach and the procedure is published as such.

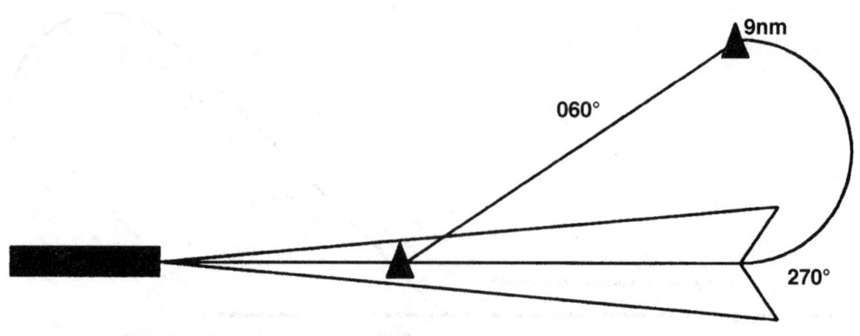

Figure 9.12 Teardrop base turn procedure.

1. The outbound leg of the hold can be extended to a predetermined distance or time.

Note: Maintain the track by using twice the drift for 1 minute and 30 seconds and then by single drift if required. At the end of the outbound leg, the QDM should be approximately ±15 degrees from the inbound axis.

2. Descend on the outbound leg as per the procedure.
3. Commence a rate 1 turn onto the final approach track.

Be prepared during an interview for the fact that you may be given some practical IFR questions; e.g., "If you are flying a heading of 076 degrees to track 088 degrees, what heading is required to maintain the reciprocal track?" The only way to prepare for these questions is to be fully conversant with instrument flight rules and procedures.

Radiotelephony

What are the radiotelephony emergency frequencies?

121.50 and 243.00 VHF.

What are the International Civil Aviation Organization (ICAO) transponder special codes?

0000	Transponder mode C malfunction, i.e., a 200-ft or greater difference between the aircraft and ATCs radar height readings if you cannot separately turn off mode C.
7700	Emergency
7600	Two-way communications failure
7500	Unlawful interference, hijacking, or unlawful interception

Note: Some local authorities may advise that 7700 be selected in the event of an unlawful airborne interception.

7000 VFR conspicuity code (1200 in the United States)

2000 IFR conspicuity code

0033 Aircraft engaged in parachuting operations

What information and format must be included in an instrument flight rules (IFR) position report?

An aircraft under IFR that flies in or intends to enter controlled airspace must report its time, position, and level at such reporting points or at such intervals of time as may be notified or directed by air traffic control (ATC).

A typical position report is:

(Aircraft identification) November four six five foxtrot tango

(Position and time) Haven, two six

(Level) flight level three five zero

(Next position and estimate) Key West, five eight

What is a Mayday call, and when is it used?

The radiotelephony (RT) spoken word *Mayday* indicates that an aircraft is threatened by grave and immediate danger and requests immediate assistance. A typical Mayday call from a pilot would be

1. Mayday, Mayday, Mayday
2. ATC name, e.g., Atlanta control
3. Aircraft call sign
4. Message, including relevant points, i.e., failures, actions being taken, and requirements.

What is a Panpan call, and when is it used?

The radiotelephony (RT) spoken word *panpan* indicates that the commander of an aircraft has an urgent message to transmit concerning the safety of his or her aircraft or person(s) on board or another aircraft, person, or property within sight. A typical Panpan call from a pilot would be

1. Panpan, Panpan
2. ATC name, e.g., Atlanta control
3. Aircraft call sign
4. Message, including relevant points, i.e., failures, actions being taken, and requirements

What is the procedure for a two-way communications failure?

The basic procedure for a two-way communications failure or a reception failure that is assumed a two-way failure is as follows:

1. Continue your flight in accordance with your current flight plan. Maintain your last cruising level given by air traffic control (ATC), or if no level was assigned, then maintain the flight plan level. Select the 7600 transponder code. Continue routine position reports in case your transmissions are still being received by ATC.
2. Arrange your flight to arrive over the arrival holding point as close as possible to your last acknowledged ETA with ATC. Or if you had no ATC-acknowledged ETA, then plan to arrive at the holding point at the calculated ETA computed from the last successful position report and flight plan data. Follow appropriate routes, i.e., STARs, etc.
3. Commence your descent within 10 minutes from over the holding fix at the last acknowledged EAT (estimated approach time) or the calculated ETA.

Note: If unable to start the descent within 10 minutes of the EAT, then do not attempt a landing, and proceed immediately to an alternate. Descend to the lowest level in the holding stack at not less than 500 ft/min, and plan to land within 30 minutes of the time your descent should have started.

4. If you are unable to complete the approach within 30 minutes of the time your descent should have started or you do not become visual at your MDA(H), then you should leave the area of the aerodrome and controlled airspace at your specified route and altitude to your alternate. Or if no altitude or route was specified, fly at either the last assigned altitude or the minimum sector safe altitude, and avoid areas of dense traffic.

Note: If you have to carry out a missed approach within the elapsed 30-minute period, then you can continue.

5. Inform ATC ASAP after landing at your destination or alternate airfield.

Air Law

When flying over or even landing at a foreign state, whose air law do you abide by?

You must obey the law of the country over which you are flying or in which you are landing so that your aircraft is not used for a purpose that is prejudicial to the security, public order, health of, or safety of air navigation in relation to that country. However, whenever your own country's, i.e., the country of the aircraft's registration, legislation (on any particular issue) is more limiting, then this should take precedence and should be adhered to.

Whose ultimate responsibility is it to avoid any type of collision?

Regardless of any air traffic control (ATC) clearance, it is the duty of the commander of an aircraft to take all possible measures to see that he or she does not collide with any other aircraft or terrain.

An aircraft must not fly so close to any other aircraft or terrain that it creates a collision danger. An aircraft that is obliged to give way to another aircraft must avoid passing over or under or crossing ahead of the other aircraft unless it is passing well clear of the other aircraft.

Note: An aircraft with the right of way should maintain its course and speed.

What does a constant relative bearing of another aircraft at the same altitude mean?

This means that a collision risk exists. (*See Q: Describe an aircraft's navigation and anticollision light, page 309.*)

With two aircraft converging in the air, which one must give way, and how?

When two aircraft are converging, i.e., at a constant relative bearing at the same altitude, the aircraft that has the other on its right must give way by turning right and passing behind the opposing aircraft.

An aircraft in the air must give way to other converging aircraft regardless of position as follows:

1. Powered aircraft must give way to those towing objects such as banners.
2. Flying machines must give way to airships, gliders, and balloons.
3. Airships must give way to gliders and balloons.
4. Gliders must give way to balloons.

What actions should be taken if you are approaching head on with another aircraft at the same altitude?

When two aircraft are approximately approaching head on at the same altitude, there is a definite danger of a collision. Therefore, both aircraft must turn right to avoid collision.

How would you overtake (in the air) another aircraft at the same altitude and direction of flight?

Whether climbing, descending, or in level flight, an aircraft overtaking another aircraft must turn right and keep right of the other aircraft during the overtaking maneuver.

Who has the right of way during an airborne overtaking maneuver?

The aircraft being overtaken in the air has the right of way.

Who has the right of way on the ground?

The right of way on the ground is as follows:

1. Regardless of air traffic control (ATC) clearance, it is the duty of the aircraft commander to do all that is possible to avoid a collision with other aircraft or vehicles on the ground. (*See Q: What is the order of priority for all vehicles on an aerodrome? page 308.*)
2. Aircraft on the ground must give way to those taking off or landing and to any vehicle towing an aircraft.
3. When two aircraft are approaching head on, each must turn right to avoid the other.
4. When two aircraft are converging, the one that has the other on its right must give way either by stopping or turning to pass behind the other. Avoid crossing in front of the other unless passing well clear.
5. An aircraft that is being overtaken by another has the right of way, and the overtaking aircraft must keep out of the way by turning left until past and well clear.

Note: This enables the commander of the overtaken aircraft (who has the right of way) to have an unrestricted view of the overtaking aircraft.

What is the order of priority for all vehicles on an aerodrome?

The priority for all vehicles on an aerodrome is as follows:

1. Aircraft landing
2. Aircraft taking off
3. Vehicles towing aircraft
4. Aircraft taxiing
5. Other vehicles

Who has the right of way between a landing aircraft and an aircraft on the ground?

An aircraft landing or on final approach and cleared to land has the right of way over other aircraft in flight or on the ground.

Note: In the case of two or more aircraft approaching to land, the lower aircraft has the right of way, except when the commander of the lower aircraft is aware that another aircraft is making an emergency landing. Then he or she must give way, and at night, even if he or she has already been cleared to land, he or she must not attempt to land until given further permission from air traffic control (ATC).

Describe the aircraft's navigation and anticollision lights.

An aircraft basically has the following navigation and anticollision lights:

1. A steady green navigation light on the starboard (right) wing that shows from dead ahead to around 110 degrees on the right side.
2. A steady red navigation light on the port (left) wing that shows from dead ahead to around 110 degrees on the left side.
3. A steady white navigation tail light that shows through 70 degrees on either side of dead astern.

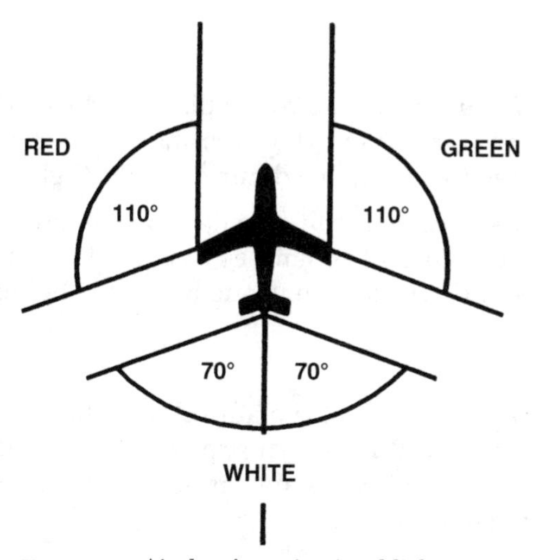

Figure 9.13 Airplane's navigational lights.

4. A flashing red anticollision light showing in all directions.
5. High-intensity white strobe lights positioned on both wing tips are found on most commercial aircraft. These are used to make the aircraft more visible to other aircraft.

When should the anticollision and navigation lights be switched on?

Anticollision light. The red anticollision light should be switched on whenever an engine is running by day or by night. The white strobe lights should be switched on whenever the aircraft is airborne. Many aircraft have a switch mode that automatically switches the strobe lights on when airborne.

Note: However, it is good airmanship to switch the strobe lights on whenever the aircraft enters or crosses an active runway. It is permissible to switch off or reduce the intensity of the flashing strobe lights if they become a distraction to crew, especially in clouds, or to persons on the ground.

Navigation lights. The green and red navigation lights should be switched on during flight at night.

Note: However, it is good airmanship to display the navigation lights during the daytime.

What actions should be taken for failed navigation and anticollision lights?

Notwithstanding any individual company minimum equipment list (MEL), the following are generally the limits imposed by most local authorities:
At night:

On the ground. Failure of the anticollision light or the navigation lights on the ground means the aircraft cannot dispatch because they are required to be displayed during night flights.

In flight. Failure of the anticollision light or the navigation lights in flight requires the aircraft to land as soon as it is safe and possible to do so unless authorized to continue by air traffic control (ATC).

During the day:

Ground and flight. Failure of the anticollision light during daylight does not stop an aircraft from flying, provided that the light is fixed at the earliest opportunity. The navigation lights are not required to be on during daylight hours.

Describe the aeronautical light beacons at airfields.

Aeronautical light beacons are installed at various civil and military airfields in some countries. Their hours of operation vary, but generally are the lights on at night and by day in bad visibility whenever the airfield is in operation.

Identification beacon. Flashes a two-letter morse identifier for the aerodrome.

Note: Usually green at civil aerodromes and red at military aerodromes. They allow a visual identification as well as a bearing to the aerodrome.

Aerodrome beacon. Flashes an alternating signal as a homing signal to the aerodrome.

Note: Usually white/white or, less commonly white/green. They are not normally provided in addition to an identification beacon.

Describe a typical approach lighting arrangement for an instrument landing system (ILS) runway.

A typical approach lighting arrangement for an ILS-equipped runway would consist of

1. Extended centerline white lights with
2. Up to three crossbars sited at 300-m intervals back along the approach path from the threshold.

Note: For increased approach lighting, more crossbars can be used, which may be sited at 150-m intervals.

Note: Approach lights simply act as a lead-in to a runway and do not mark the boundaries of a suitable landing area.

In addition, an advanced precision approach (CAT II and III) lighting arrangement could be never ending, but as a minimum you could expect

3. An all-round increase in white approach centerline and crossbar lights at reduced spacing intervals and higher intensity.
4. High-intensity red approach lights on either side of the centerline white approach lights and in between the white crossbars. In addition, these lights can be found on the runway prior to the touchdown zone.

Note: These would be listed on the runway approach plate.

5. Extra green threshold wing bar lights that are extended outboard of the runway edge threshold lights or at the displaced threshold.

Describe a typical runway lighting arrangement.

Runway centerline lights. Precision approach runways usually have flush white runway centerline lights. The color of these lights changes over the last 600 m (caution zone) to alternate red and white lights. Then over the last 300 m they change to all red lights.

However, some runways, especially nonprecision approach runways, have no centerline lights, just edge lighting.

Touchdown zone lights. Consisting of rows of flush white lights on either side of the runway centerline during the first 900 m. Commonly used for CAT II and III runways.

Runway edge lights. White lights that may be directional only lights, i.e., only visible to aircraft aligned with the runway, or they may be equally visible in all directions. Some are flush with the runway, and others are elevated. For precision approach runways, the color of the runway edge lights changes over the last 600 m (caution zone) to alternate red and white (sometimes yellow) lights. Then over the last 300 m they change to all red lights.

Threshold lights. Green threshold lights cross the width of the runway threshold.

Note: Extra green threshold wing bar lights are sometimes extended outboard of the runway edge threshold lights or at the displaced threshold for poor-visibility approaches.

What do you need to see to continue at the minimum decision height (MDH) for a nonprecision approach?

The approach may not be continued below MDH(A) unless at least one of the following visual references for the intended runway is distinctly visible to and identifiable by the pilot:

1. Element of the approach light system
2. The threshold
3. The threshold markings
4. The threshold lights
5. The threshold identification lights
6. The visual glide slope indicator
7. The touchdown zone lights
8. The touchdown zone or touchdown zone markings
9. The runway edge lights

What do you need to see to continue at decision height (DH) for a CAT I approach?

The approach may not be continued below CAT I DH(A) unless at least one of the following visual references for the intended runway is distinctly visible to and identifiable by the pilot:

1. Element of the approach light system
2. The threshold
3. The threshold markings
4. The threshold lights
5. The threshold identification lights
6. The visual glide slope indicator
7. The touchdown zone lights
8. The touchdown zone or touchdown zone markings
9. The runway edge lights

What do you need to see to continue at decision height (DH) for a CAT II approach?

The approach may not be continued below CAT II DH unless there is a visual reference containing a segment of at least three consecutive lights:

- The centerline of the approach lights, or
- The touchdown zone lights, or
- The runway centerline lights, or
- The runway edge lights or a combination of these

What do you need to see to continue at decision height (DH) for a CAT IIIa approach?

The approach may not be continued below CAT IIIa DH unless there is a visual reference containing a segment of at least three consecutive lights:

- The centerline of the approach lights, or
- The touchdown zone lights, or
- The runway centerline lights or edge lights, a combination of these

What are VASI lights?

Visual approach slope indicator (VASI) lights are a system used to provide slope guidance during the visual stage of the approach. They will assist pilots to maintain a stable descent path down to the runway surface for the flare and landing.

Two-bar VASI. A typical two-bar VASI system has two wing bar single lights, one above the other, which are positioned at the side of the runway, usually at 300 m from the approach threshold. A duplicated set is positioned on the opposite side of the runway. This is sometimes known as the red-white system because the colors seen by the pilot tell him or her if he or she is right on slope, too high, or too low as follows:

- All bars white = high on the slope
- Near bars white and the far bars red = right on slope
- All bars red = low on the slope

Three-bar VASI. This has an extra wing bar single light further down the runway to assist pilots of larger aircraft, such as the B747. This is so because the guidance given by VASIs depends on the position of the pilot's eyes. Because the wheels of a large aircraft are much further below the pilot's eyes, it is essential that his or her eyes follow a parallel but higher slope to ensure adequate clearance over the runway threshold. Pilots of large aircraft should use the second and third wing bar lights and ignore the first, whereas pilots of smaller aircraft should use the two nearer wing bar lights and ignore the third furthest. The red-white guidance system is the same as the two-bar VASI system.

What are PAPI lights?

Precision approach path indicator (PAPI) lights are a development of the visual approach slope indicator (VASI) system that also uses the red-white signals for guidance to maintain the correct approach path. However, the lights are arranged differently. The PAPI system has a single wing bar that consists of four light units on one or both sides of the runway adjacent to the touchdown point. The colors seen by the pilot tell him or her if he or she is right on slope, too high, or too low.

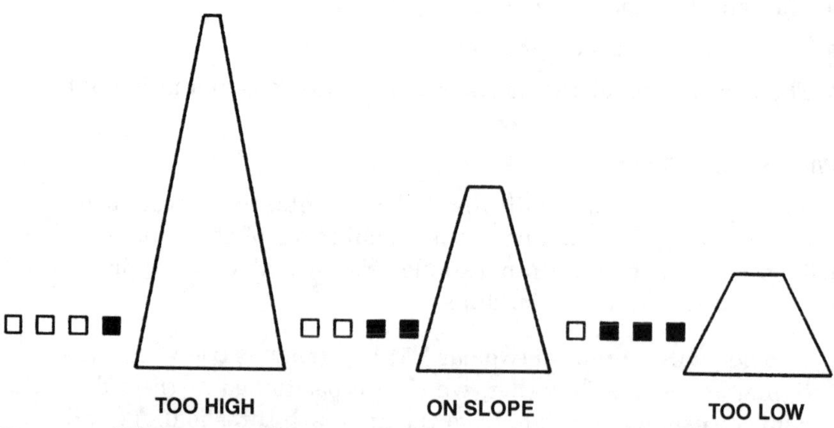

Figure 9.14 PAPI slope guidance.

What does alternate red and white edge/centerline lights on a runway indicate?

The last 600 m (2000 ft) of the runway for U.S. ASLF runway lights or the last 900 m (3000 ft) of the runway for British Calvert-style lighting systems.

What does red only edge/centerline lights on a runway indicate?

The last 300 m (1000 ft) of the runway.

Describe the two types of taxiway lighting systems.

The two types of taxiway lighting systems are

1. One line of green taxiway centerline lights
2. Two lines of blue taxiway edge lights

Note: At certain points on the taxiway there may be red stop bars to indicate the position of a hold, for instance, before entering an active runway.

What does two or more white crosses displayed on a runway or taxiway indicate?

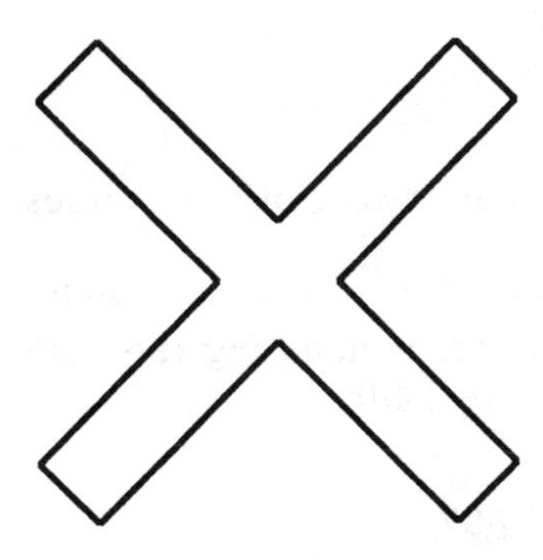

Figure 9.15 White cross markings on a taxiway or a runway.

White crosses displayed on a runway or taxiway indicate that the surface is unfit for movement of aircraft, namely, that the area is closed.

Note: Taxiway markings may be yellow. However, the surface may be used in emergencies, but be aware that the runway or taxiway may have obstructions or be used for storage.

Describe the usual runway holding markings.

A normal CAT I hold is marked by two close parallel yellow lines across the taxiway (old markings may still be white), one continuous line and one broken line. It signifies a holding position beyond which no part of an aircraft or vehicle may project across in the direction of the runway without air traffic control (ATC) permission. The CAT I hold is usually the last holding point prior to the runway.

Note: The pilot can determine if the hold affects him or her, i.e., if you encounter the continuous line first and then the hold affects you (you are on the taxiway holding side) and you need ATC clearance to cross. Moving in the reverse direction, where the broken line is encountered first, i.e., runway side, the holding point does not affect you, and you do not require a clearance to cross it.

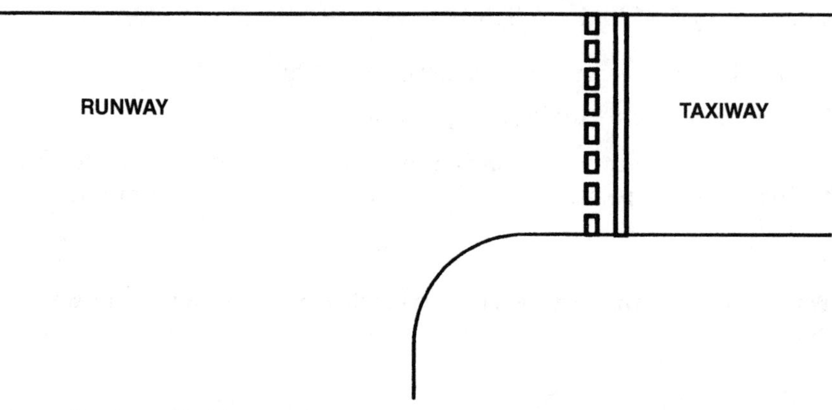

Figure 9.16 Normal holding point.

Note: Some runway hold markings may have two continuous lines and two broken parallel lines.

In addition, a runway holding point usually will have a red and white sign on either side of the taxiway hold markings denoting the hold category, i.e., CAT I hold, and runway, i.e., Rwy 24/06.

Describe the CAT II and III runway holding markings.

A CAT II or III holding point has a yellow ladder-type marking. A CAT II or III holding point is usually further back from the runway than a CAT I hold (CAT III hold is further back than a CAT II hold) yet still has the same restrictions as the CAT I hold when in force. (*See Figure 9.17*)

Should you report any hazardous flight conditions?

The commander of an aircraft must report to air traffic control (ATC) as soon as possible any hazardous flight conditions encountered.

Note: Typical conditions worthy of reporting are moderate or severe turbulence, windshear, volcanic ash, and rapidly deteriorating visibility.

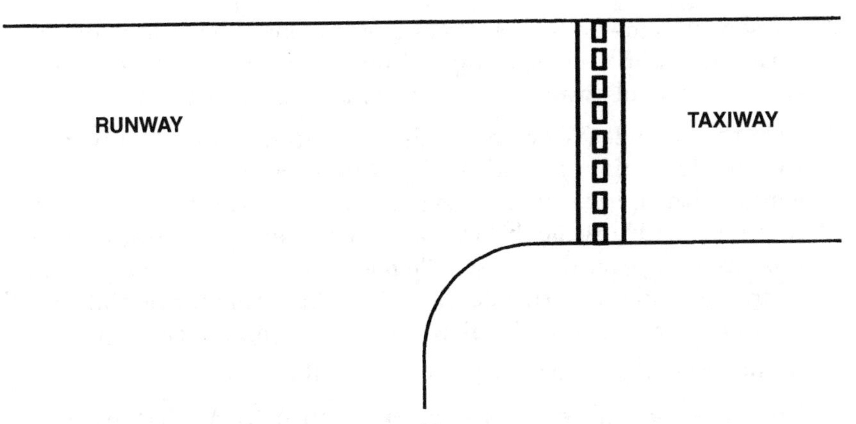

Figure 9.17 CAT II or III holding point markings.

When must the pilots be at the aircraft's controls?

The commander of aircraft is responsible for ensuring that

1. One pilot is at the controls at all times in flight.
2. Both pilots are at the controls during takeoff and landing if the aircraft is required to carry two pilots. This usually means during the climb to 10,000 ft or transition, whichever is the latter, and from the top of descent.

Note: Each pilot at the controls must wear a safety harness during takeoff and landing where one is provided and required and a lap safety belt at all other times.

Adverse Weather Recognition and Flight Technique

Describe flight technique in moderate/severe weather/turbulent rough air.

First, avoid areas of severe weather and turbulence:

1. Do not take off or land in areas of *severe* weather and turbulence.
2. In flight, avoid predicted areas of turbulence by
   a. Flight planning adjustments if possible, namely, changing routing
   b. Using the weather radar, monitoring weather reporting stations, in-flight reports on air traffic control (ATC) frequencies, and visual observations

Second, if committed to areas of moderate/severe weather in flight, adopt the following technique:

1. Ignition on/autopilot on

2. Passenger signs on (seat belts and no smoking). In thunderstorms, turn the flight deck lighting full on, night or day. Stow all loose equipment, and strap yourself in tight using the full harness.
3. Hold the attitude (i.e., climb, cruise, or descent). Note, however, that you should sacrifice altitude to maintain attitude. However, be fairly gentle with the use of the elevator to avoid overstressing the airframe structure through large changes in angle of attack and lift produced. Also, for this reason, do not try to climb over a storm from a straight and level attitude. Check height for maneuverability and terrain clearance. Use the ailerons to maintain lateral control.
4. Autothrust off to protect against engine flameout.
5. Select the rough air speed (*see Q: What is VRA/MRA speed? page 208*), which may require reducing power. Allow the airspeed to fluctuate more than usual, but be prepared to correct for any large speed diversions by adjusting the power.

Further:

1. Continue to apply avoidance techniques.
   a. Monitor the weather radar for the best passage through the severe weather.
   b. Use the deicing systems as required, and ensure that the pitot heat is on.
2. Be prepared for other disturbances.
   a. High levels of static on some radios may necessitate that they are turned off.
   b. Aircraft electrical systems and compasses may be affected temporarily by strong local electrical fields. Therefore, check all flight instruments and supplies.
   c. You may get sparks or small electrical shocks if the aircraft is struck by lightning, or the whole aircraft may be surrounded by St Elmo's fire.
   d. You may hear loud battering noises if you fly into hail.

How do you use weather radar information?

Weather radar must be used as a means to circumnavigate (avoid) storms by turning upwind where possible around a storm cell.

Note: A rule of thumb would be below 30,000 ft, avoid cells by approximately 10 to 15 nautical miles; above 30,000 ft, avoid cells by approximately 15 to 20 nautical miles.

Do not use the weather radar to penetrate severe storms by making small changes through the middle of a cluster of storm cells. This is so because severe windshear can be experienced in between cells, and you may be led up a blind alley due to one cell shielding another from your radar. (*See Q: How does an isoecho/color radar work? page 112.*)

What effect does windshear have during an approach for landing?

A change in wind direction and/or speed windshear has the following effects on an aircraft during the landing approach:

1. If the headwind component increases, then the aircraft will experience a transient increase in performance. This will lead to an overshoot tendency.
2. If the headwind component decreases, then the aircraft will experience a transient decrease in performance.

Note: This is a typical scenario as you get closer to the ground, causing an increase in the rate of descent if allowed to go unchecked. However, normally, this effect is only slight and doesn't cause too much of a concern, although any significant windshear drop should be treated with respect.

Loss of performance in this scenario will cause the aircraft to sink and possibly undershoot the aiming point. (*See Qs: What is windshear? page 254; How does windshear affect an aircraft? page 255.*)

For the effect of severe windshear (microbursts) on an aircraft during the landing approach, see Q: How does a microburst affect an aircraft? page 258.

What indications should you look for if windshear is expected?

You should look for indicated airspeed, temperature, and lift trends. That is, an increasing headwind would increase the indicated airspeed and lift performance for no power or attitude changes. And a decrease in the headwind would decrease the indicated airspeed and lift performance for no power or attitude changes. The rate at which these changes occur indicates the severity of the windshear ahead. (*See Q: How is windshear detected? page 255.*)

How would you fly an approach if you suspect windshear?

You would increase the approach speed to compensate for the loss of energy that is common with low-level windshear. This action will help to guard against a stall when you attempt to maintain the flight path by increasing the aircraft's attitude. (*See Qs: Where do you find windshear? page 254; What is the recovery technique for windshear? page 320.*)

You would monitor the conditions ahead in case the conditions worsened from mild to severe windshear (microbursts).

What is the recovery technique for windshear?

For mild or even moderate windshear that causes a manageable flight path deviation, i.e., less than a single dot deviation on the approach, then it is acceptable to use normal power (for speed) and pitch (for the descent profile) corrections to regain the desired flight path.

Note: Unless it is believed that this condition is a warning of more severe disturbance ahead. If this is the case, then you should adopt a more aggressive attitude and initiate an immediate go-around.

For severe windshear (which can include microbursts), a TOGA full-power and maximum pitch-up go-around should be initiated immediately. (*See Q: How would you fly an approach if you suspected a microburst? page 320.*)

You are flying an approach and you experience rain on the windscreen. What should you be cautious of, or what is the biggest hazard in heavy rain on final approach?

Microburst. (*See Qs: What is a microburst? page 256; Where do you find microbursts? page 256; and What do you know about microbursts? page 256.*)

How would you fly an approach if you suspected a microburst?

You should not attempt an approach into an area where a microburst is reported or likely. (*See Q: Where do you find microbursts? page 256.*) Hold for 10 to 15 minutes to allow the storm to move away from the approach path before attempting an approach. Continue to monitor for indications of possible microburst activity. (*See Q: What do you know about microbursts? page 256.*)

Note: The early identification of a possible microburst or severe windshear downdrafts on an approach is vital to ensure that you have sufficient power, height, and speed reserves available to successfully initiate a go-around. (*See Q: What identifies (are the indications of) a microburst? page 257.*) Therefore, it is important not to overcompensate for an overshoot tendency by reducing power early in an approach because this will incur a delay in the delivery of full power when you pass into the downdraft area later on the approach and when the go-around has to be initiated. This time delay could be crucial. (*See Q: How does a microburst affect an aircraft? page 258.*)

If, however, you do encounter an unexpected microburst on an approach, you should initiate an immediate full-power (TOGA) maximum pitch-up go-around. If you are close to the ground and the landing gear is already down, then you should delay retracting the gear until ground contact is definitely no longer a possibility.

What is aquaplaning/hydroplaning?

Aquaplaning is the phenomenon of a tire skating (not rotating) over the runway surface on a thin film water. It is caused by the buildup of a layer of water beneath the tire, which becomes a wedge between the tire and the runway. This wedge of water effectively lifts the tire off the ground, reducing the friction forces to practically zero. Therefore, wheel braking has no effect, and directional control also may be lost. Aquaplaning is extremely significant on jet transport aircraft and can increase the stopping distance required significantly.

What are the three types of aquaplaning/hydroplaning?

The three types of aquaplaning are

1. *Dynamic.* This is due to standing water on the runway when the tire is lifted off and completely supported. That is, it is a function of fluid density.
2. *Viscous.* This occurs when the surface is damp and provides a very thin film of fluid that cannot be penetrated by the tire. Viscous aquaplaning can occur at or persists down to much lower speeds than simple dynamic aquaplaning. It is particularly associated with smooth surfaces and is quite likely to occur in the touchdown area, which is often liberally smeared with rubber deposits. That is, it is a function of fluid viscosity.
3. *Reverted rubber or rubber reversion.* This refers to the tire becoming tacky. It requires a long skid, rubber reversion, and a wet surface. The heat from the friction between the tire and the wet runway surface boils the water and reverts the rubber, which forms a seal that delays water dispersal. The steam then prevents the tire from contacting the runway. That is, melting rubber traps the steam, which causes aquaplaning.

Note: All three types of aquaplaning can occur during one landing run if the conditions are correct.

How do you calculate the aquaplaning/hydroplaning speed?

The approximate minimum true ground speed in knots at which aquaplaning can be *initiated,* is calculated using the following formula

9 × square root of the tire pressure in pounds per square inch for takeoff

7.6 × square root of the tire pressure in pounds per square inch for landing

However, it is generally regarded that 9 times the square root be used also for the landing, although in theory only 7.6 times the square root is required at landing, followed by 9 times the square root during the rollout.

Note: Aquaplaning occurs at any speed above this minimum speed. On takeoff, there is normally a considerable gap, i.e., 20 to 50 knots, between the minimum aquaplane speed and V_1, e.g., minimum aquaplane speed of 115 knots and a V_1 speed of 145 knots. On landing or a rejected takeoff, if aquaplaning occurs, it may continue at speeds lower than the calculated aquaplane speed. This is so because the calculation gives us a speed below which aquaplaning will not *start.*

Tire wear is obviously an important factor; the more worn the tire, the more likely it will be to aquaplane. A general awareness of this should be retained.

How do you control an aquaplane?

You control an aquaplane by using the antiskid braking systems, which releases the brakes if it senses a skid. This allows the wheel to rotate and to push through the water layer, making contact with the runway again.

If you have no antiskid systems or they are inoperative, then manually brake using the same principle, namely, if you feel a slip, then release the brakes, allow the skid to stop, and then reapply the brakes.

Initially on a wet runway, what is the most effective means of stopping?

Thrust reverses. (*See Q: What is the most efficient system for stopping at high speed? page 287.*)

What considerations should you take for taking off in icing conditions?

Your considerations for taking off from a contaminated runway with snow or slush in icing conditions starts from your preflight preparations through the startup, taxi, actual takeoff, and the after-takeoff phases of flight. (*See Q: Describe the in-flight precautions for icing conditions, page 324.*)

Operations from a snow- or slush-covered runway involve a significant element of risk. Therefore, operations from contaminated runways should be avoided wherever possible. However, if you are committed to a departure in these conditions, then consider the following:

1. Deice the aircraft either on standby prior to engine start or by taxiing through a deicing station.

2. Ensure that engine anti-ice is switched on during ground operations in icing conditions. Use carburetor heat and propeller deicing if appropriate.

Note: In the absence of guidance from your aircraft's flight manual, engine icing is considered possible if the outside air temperature (OAT) is +10°C or less with visible moisture. (*See Q: In what conditions do you expect icing conditions, and should you turn on the engine anti-icing? page 324.*)

3. During taxiing in icing conditions, the use of reverse thrust on podded engines should be avoided because this can result in ice contamination of the wing leading edges and slats. For the same reason, keep a good distance behind the aircraft ahead. In no circumstances should an attempt be made to deice an aircraft by positioning it in the wake of the engine exhaust of another aircraft.
4. Before takeoff, ensure that the wings are not contaminated by ice or snow. Operate all the appropriate anti- and deicing systems, e.g., engine, wing, propeller, carburetor, probes, etc., before and after takeoff in accordance with the aircraft's flight manual. Takeoff power should be monitored on more than one instrument, e.g., EPR and N_1 gauges. This is in case of icing on any engine probes that causes an overreading on the main engine power instrument. Monitoring a second instrument can cross-check the reliability of the main instrument.

What are the two main types of anti-icing fluids used for deicing on the ground?

1. *Type I fluids (unthickened).* These fluids have a high glycol content and a low viscosity. The deicing performance is good, but they provide only limited protection against refreezing.
2. *Type II fluids (thickened).* These fluids have a minimum glycol content of approximately 50 percent and, due to the thickening agent, have special properties that enable the fluid to remain on the aircraft surfaces, known as *holdover time.* The deicing performance is good, and in addition, protection is provided against refreezing and/or buildup of further accretions when exposed to freezing conditions. Therefore, they are also anti-icing fluids. Their anti-ice holdover time depends on the following factors:
   a. Type of snow
   b. Wet or dry snow (Note that wet snow will result in a shorter holdover time)
   c. Airframe temperature

d. Outside air temperature (OAT)
e. Amount of precipitation

The deicing fluid applied would be described as follows: Type of fluid, i.e., I or II percent of fluid to water mixture, holdover time expected.

Describe the in-flight precautions for icing conditions.

Icing during flight can occur at any time of the year. Therefore, a pilot should take the following precautions when icing conditions are possible:

1. Pilots should avoid icing conditions for which the aircraft is not approved.
2. Keep probe heating on when airborne.
3. Crews should visually check for the buildup of ice on the airframe at regular intervals and select wing anti-icing when required.
4. Crews should monitor the air temperature and signs of visible moisture regularly and select engine anti-icing when required.

Note: Switch on and off the engine anti-icing systems separately to guard against simultaneous engine flameouts from ice ingestion.

5. Be aware of the influence of ice buildup on the performance and controllability.
   a. The performance effect of wing contamination includes
      (1) Reduced stalling angle of attack
      (2) Increased stalling speed
      (3) Reduced maximum lift capability
      (4) Reduced amount of lift at a given angle of attack, especially at higher angles of attack
   b. Controllability difficulties of the aircraft due to ice buildup can be manifested in many different situations. For example, the deployment of flaps may cause a reduction in longitudinal controllability when ice is present on the tailplane. When this happens, the tailplane can stall with subsequent loss of control, from which recovery may be impossible in the time/height available. Therefore, if tailplane icing is suspected, then it would be wise not to deploy flaps. Also be aware of a lateral imbalance if you were to only deice one wing; on some aircraft types, this can occur with an engine failure and the cross-bleed closed.

In what conditions would you expect icing, and when should you turn on the engine anti-icing?

You should expect icing in the following conditions and therefore turn on the engine anti-icing systems.

1. *On the ground.* When the outside air temperature (OAT) is +10°C or lower with visible moisture present.

Note: Static air temperature (SAT) measurements usually are considered to represent ground OAT.

2. *In flight.*
   a. *During the climb and cruise.* When the total air temperature (TAT), which is SAT plus the kinetic effect of the aircraft, is colder than +10°C but warmer than −40°C, with visible moisture present.

Note: You should not turn on the engine anti-icing systems in the climb or cruise when the SAT is colder than −40°C. This is so because only ice crystals are present in the atmosphere at these extreme temperatures. These crystals do not pose an icing threat because they have no liquid state that can stick to the aircraft.

   b. *Descent.* When the TAT is colder than +10°C with visible moisture present.

Note: Even with a SAT colder than −40°C. This is so because in a descent the temperature will increase, and the SAT will become warmer than −40°C, creating a real risk of supercooled water droplet icing. Therefore, the engine anti-icing should be switched on at the top of descent whenever the TAT is colder than +10°C and visible moisture is present.

When would you expect carburetor icing in a piston engine?

Carburetor icing should be expected in a piston engine when the outside air temperature (OAT) is between −10°C to +30°C with a high humidity and/or visible moisture present in the air. However, carburetor/throttle icing is most likely to occur between +10 and +15°C with a relative humidity greater than 40 percent.

Note: Carburetor icing *can* be found on a warm day in moist air, especially with descent power settings.

What actions should you take to prevent or remove carburetor/throttle icing?

You should use the carburetor heat system (hot air) at regular intervals to treat icing in the carburetor when in carburetor icing conditions. The carb heat system delivers hot air from the engine compartment into the carburetor that melts the buildup of ice.

Note: It is also advisable to apply carb heat at the start of a descent to protect against throttle icing occurring and thereby ensuring that full power is available in the event of a go-around engine application. However, carb heat always should be switched off before the throttle is advance to ensure maximum thrust delivery, and therefore, it should be off in the final landing approach in anticipation of an emergency go-around near to the ground.

However, the use of carburetor heat should be avoided when the OAT is colder than −10°C because by applying carburetor heat you raise the air temperature in the carburetor into the icing temperature range, i.e., −10°C to +30°C.

Adverse Conditions Flight Technique

What are the hazards associated with flying in a region of volcanic ash?

Engine flameout.

This is due to the engine being starved of air because

1. Ash builds up on fan and compressor blades, which upsets the airflow through the engine.
2. High volcanic ash content in a parcel of air also deprives the combustion chamber of the air it requires for a controlled explosion.

What procedures would you adopt when flying in a region of volcanic ash?

1. On first observation or contact with volcanic ash, make an immediate 180-degree turn away from the region of volcanic ash.
2. If inadvertently you encounter a region of volcanic ash, select all engine high-bleed systems on, e.g., air conditioning, engine and airframe deicing/anti-icing systems, etc.

Note: This helps to reduce the high volume of ash from the engine, thereby helping to maintain a higher percentage of air delivery to the combustion chambers to protect against a flameout.

3. Auxiliary power unit (APU) on to assist engine relight if necessary.

In addition, you should apply any further procedures that your flight manual may recommend.

Chapter

10

Human Performance

What should your sleep pattern be before a flight?

A pilot should manage his or her sleep pattern so that he or she awakes from a sleep a few hours before he or she is to report for duty so that he or she arrives for the flight fresh and awake as if it where the beginning of a new day at the office. Obviously, this is difficult for commercial pilots, especially long-haul pilots, who have to sleep at times when their bodies are saying they should be awake.

What causes fatigue?

Fatigue is caused by (1) the lack of sleep and (2) overwork/stress.

What are the consequences of fatigue?

Fatigue causes

1. An increase in a pilot's response time
2. Impaired judgment
3. An increase in the possibility of human error

What is hypoxia?

Hypoxia is a human condition that occurs when the oxygen supply available to the human tissues is insufficient to meet their needs, especially respiratory requirements. Hypoxia means a lack of oxygen.

What causes hypoxia?

Hypoxia is caused by altitude. The greater the altitude, the greater is the hypoxia, and the more rapid is its progression. This is so because respiration brings less oxygen into the body at high altitudes because of the decrease in the ambient air pressure/density that produces a corresponding fall in the partial pressure of oxygen in the atmosphere. (*See Q: How long do you have before you become unconscious from hypoxia at _____ height? page 328.*)

Susceptibility to hypoxia includes the following effects:

1. *Time.* The longer the exposure time to a lack of oxygen, the greater is the effect of hypoxia.
2. *Exercise.* This increases the demand for oxygen and hence increases the degree of hypoxia.
3. *Illness.* This increases the energy demand of the body and therefore oxygen demand, and thus this causes an increase in the degree of hypoxia experienced.
4. *Fatigue.* This lowers the threshold for hypoxia symptoms.
5. *Drugs and alcohol.* These reduce the body's tolerance of altitude and therefore increase an individual's susceptibility to hypoxia.
6. *Smoking.* Carbon monoxide produced by smoking binds to hemoglobin with a far greater affinity than oxygen and therefore has the effect of reducing the hemoglobin for oxygen transport, exacerbating any degree of hypoxia.

How long do you have before you become unconscious from hypoxia at _____ height?

Individuals' time of useful consciousness with hypoxia is as follows:

18,000 ft	30 minutes
25,000 ft	2–3 minutes
30,000 ft	45–74 seconds
40,000 ft	20 seconds
45,000 ft	12 seconds

What are the indications/symptoms of hypoxia?

The indications/symptoms of hypoxia are

1. Cyanosis (This the development of a blue color in the parts of the body where the blood supply is close to the surface, e.g., lips and fingernails.)
2. Apparent personality change
3. Impaired judgment
4. Muscular impairment

5. Memory impairment
6. Sensory loss
7. Hyperventilation
8. Impairment of consciousness

The degree of these indications/symptoms varies with altitude because of the different levels of oxygen partial pressure. Therefore, a pilot should be aware of these indications/symptoms developing slowly at low altitudes and very quickly at high altitudes.

How do you treat hypoxia?

Increasing the individual's oxygen supply should treat hypoxia. Therefore, pilots should be familiar with the appropriate oxygen drills whenever they are flying at an altitude where hypoxia can occur, i.e., above 10,000 ft.

What is hyperventilation?

Hyperventilation is the condition of overbreathing, i.e., in excess of the ventilation level required to remove carbon dioxide, causing changes in the acid-base balance of the body. Hyperventilation means an excess of oxygen.

What causes hyperventilation?

Hyperventilation is caused by

1. Hypoxia
2. Anxiety
3. Motion sickness
4. Vibration
5. Heat
6. High *g* force
7. Plus many other causes to a lesser extent

What are the indications/symptoms of hyperventilation?

The indications/symptoms of hyperventilation are

1. Dizziness
2. Tingling sensation, especially in hands and lips
3. Visual disturbances, i.e., tunneling or clouding

4. Hot or cold feelings
5. Anxiety, which causes a circle of cause and effect
6. Impaired performance, often dramatically reduced performance
7. Loss of consciousness (However, therefore, respiration returns to normal, and the individual recovers.)

These indications/symptoms are very similar to the indications and symptoms of hypoxia.

How do you treat hyperventilation?

Below an altitude of 10,000 ft, symptoms of hyperventilation should be diagnosed as hyperventilation and treated accordingly by regulating the rate and depth of respiration to reduce the individual's over-breathing. Normally, this is accomplished with deep breaths into a paper bag. However, some medical bodies now suggest that you simply allow the patient to hyperventilate to the point where he or she passes out, during which time his or her body will regain a normal breathing pattern.

Above an altitude of 10,000 ft, the same symptoms should be diagnosed as hypoxia and treated accordingly by giving oxygen to the individual. This is so because it is very difficult to distinguish between the effects of hypoxia and hyperventilation, and because hypoxia is possible above 10,000 ft, the worst-scenario should be assumed, which is hypoxia.

What is the most important exhaust gas that pilots should be aware of in terms of personal safety, and why?

Carbon monoxide (CO). It may enter the cabin if there are any leaks in the exhaust system when warm air from the around the engine or the exhaust manifold is used to provide cabin heating. This gas can cause headaches, breathlessness, impaired judgment, and eventually loss of consciousness. (*See Qs: What is carbon monoxide poisoning? page 330; and What are the symptoms of carbon monoxide poisoning? page 331.*)

What is carbon monoxide poisoning?

Carbon monoxide (CO) is a highly toxic colorless, odorless, and tasteless gas for which hemoglobin in the blood has an enormous affinity. The main function of hemoglobin is to transport oxygen from the lungs throughout the body. Carbon monoxide molecules breathed into the lungs along with air will attach themselves to the hemoglobin, starving the brain and body of oxygen, even though oxygen is present in the air.

How is carbon monoxide produced?

Carbon monoxide is produced during the combustion of fuel in the engine.

Note: It may enter the cockpit if there are any leaks in the exhaust system and the cabin heating system uses warm air from around the engine or the exhaust manifold as the source of heat.

What are the indications that carbon monoxide is present?

Carbon monoxide is an odorless, invisible gas that is very difficult to detect other than

1. Cherry red skin pigmentation from prolonged exposure.
2. Specific carbon monoxide indicators/systems, e.g., cards that change color or aural alarms that sound on contact with carbon monoxide.

What are the symptoms of carbon monoxide poisoning?

The symptoms of carbon monoxide poisoning may include

1. Headache
2. A slower breathing rate or breathlessness
3. Dizziness
4. Nausea
5. Deterioration in vision
6. Impaired judgment
7. Eventually loss of consciousness

What actions should you take if you suspect carbon monoxide poisoning?

If carbon monoxide is suspected,

1. Shut off cabin heat
2. Stop all smoking
3. Increase the supply of fresh air
4. Don oxygen masks if available

What is a visual whitening effect?

Whitening is a visual illusion whereby the perception of the distance to an object's surface is incorrect. That is, a close object is perceived to be far away. A white surface or light distorting a person's visual depth acuity causes this perception.

Note: This effect may have played a part in the Air New Zealand DC10 accident in Antarctica.

What is visual empty field myopia?

Empty field myopia occurs when you are flying in clouds or pitch darkness, i.e., no or very few visual references that make your eyes automatically focus to approximately 1 m of distance ahead.

What is visual blackhole effect?

Blackhole effect occurs when a runway in virtual pitch darkness (except maybe the odd aerodrome light around) and in the middle of nowhere gives the visual illusion of the aircraft being at a greater height than its actual height.

What is the nighttime visual (eye acuity) effect?

Nighttime visual effect is a form of empty field myopia when airborne and a form of blackhole effect when landing. (*See Qs: What is visual empty field myopia? page 332; What is visual blackhole effect? page 332.*) That is, the runway always appears to be further away in terms of vertical distance than it actually is; therefore, you tend to land earlier than expected, so beware.

How long can you not fly after diving, and why?

You are not allowed to fly for 12 hours after swimming/diving with compressed air. You are not allowed to fly for 24 hours after swimming/diving with compressed air to a depth of 30 ft or more. These times are general guidelines, and some authorities/companies have greater time limits, e.g., 24 and 48 hours.

The reason for these restrictions is to protect against decompression sickness.

What is risky shift?

Risky shift is the tendency for a group of people to decide on a riskier, even more dangerous decision than an individual who is part of the group would make by himself or herself.

Are you more likely to take a risk as an individual or as a group?

It is human nature that groups of people are more likely to decide to take a greater risk than an individual would. This is generally thought to be so because in a group we take confidence in other people and their decisions more so than our own.

How long should you wait before flying after consuming alcohol?

Eight hours (after consuming a small quantity of alcohol) and ensuring that all the alcohol has left your system.

Note: One unit of alcohol takes 30 minutes to leave the human system. However, some authorities and companies may have longer periods; this is so because after heavy drinking, alcohol can stay in the body for up to 30 hours.

Chapter

11

Type-Specific Questions

The interviewer is likely to ask type-specific questions from the following parameters;

1. On the aircraft type being recruited for if the applicant has previously flown this type.
2. On your present or last aircraft type flown.

Therefore, the aircraft type under question could vary between the airline's type and/or the candidate's type, and obviously, all the possible varieties of aircraft types cannot be accommodated in this book alone. It is thus left up to the reader to be fully conversant with his or her own aircraft's design, systems, engines, and operational procedures, and the following example questions are just that, examples.

Aircraft

Describe your aircraft's wing.

As a guideline, you should cover the following areas:

1. Type: i.e., straight, tapered, swept, or delta
2. Dimension:
   a. Span
   b. Mean aerodynamic chord (MAC)
   c. Turning radius
3. Airflow control, i.e., design features:
   a. Winglets
   b. Fences
   c. Vortex generators
   d. Slots, etc.

4. Speeds
   a. Indicated Mach number at which the airflow over the wing is Mach 1 and therefore when the Mach trimmer becomes active
   b. Airspeed to be flown if the Mach trimmers fails
5. Design features
   a. Trailing-edge flaps
   b. Leading-edge flaps

Explain your aircraft's anti-icing/deicing systems?

Where is the yaw damper switch on your aircraft type?

Where is the GPWS cutout switch on your aircraft type?

Where is the flow-control valve on your aircraft type?

How is the fuel stored and distributed in your aircraft?

Engines

What type of engine is on your aircraft?

What are the approximate midweight thrust settings for takeoff, maximum continuous thrust, cruise (at service ceiling), and approach?

Operations

How do you do fuel planning on your aircraft type?

How much fuel is used on your aircraft per hour and in total?

Approximately what are your aircraft's typical speeds?

VMCA?

V_1 speed?

V_R speed?

Climb speed?

Cruise speed at 3000 ft?

Cruise speed at maximum service ceiling?

VMO?

MMO?

Outline the takeoff performance on your aircraft.

How do you maintain obstacle clearance for your aircraft, i.e., what is the profile you fly?

What does cost index mean in your aircraft?

What things are considered by the FMS when setting cruise speed in ECON mode?

Index

About the Author

Gary V. Bristow is an airline pilot. His own interviewing experiences made him realize how valuable a book such as this would be in advancing a pilot's career.